The Human Dynamics of Achieving an Injury-Free Workplace

Creating an injury-free workplace demands more than policies and procedures since it requires the engagement of every employee to actively care for their own safety and that of their colleagues. This research-based teaching and learning workbook addresses the critical human dynamics that determine whether safety and health initiatives succeed or fail. Drawing from decades of psychological science, the authors present a systematic approach to building an actively-caring-for-people (AC4P) work culture that transforms safety from a compliance-driven requirement into an interdependent shared commitment.

This teaching/learning resource provides the scientific foundation and practical guidance necessary to achieve lasting safety improvements through interpersonal human dynamics. It contains ten distinct safety objectives, each supported by six evidence-based behavioral directives that readers can immediately implement in their workplace. These directives span multiple domains of psychological science, from behavioral principles of reinforcement and feedback to humanistic concepts of empathy and systems thinking. Each chapter includes explanatory text, practical illustrations, thought-provoking discussion questions, and relevant research references, covering essential topics such as behavior-based safety coaching, incident analysis, social influence strategies, stress management, and the cultivation of a Total Safety Culture. Readers will gain the knowledge and tools needed to optimize both individual and interpersonal dynamics for occupational health and safety.

The Human Dynamics of Achieving an Injury-Free Workplace: Safety Directives from Psychological Science is for safety professionals, managers, supervisors, team leaders, and anyone responsible for workplace safety and health. It is equally valuable for safety committees, training facilitators conducting workshops, and individual practitioners seeking to enhance their understanding of safety-related psychology.

AF739696

The Human Dynamics of Achieving an Injury-Free Workplace

Safety Directives from Psychological Science

E. Scott Geller and Krista S. Geller

CRC Press
Taylor & Francis Group
Boca Raton London New York

CRC Press is an imprint of the
Taylor & Francis Group, an **informa** business

Designed cover image: E. Scott Geller and Krista S. Geller

First edition published 2027
by CRC Press
2385 NW Executive Center Drive, Suite 320, Boca Raton FL 33431

and by CRC Press
4 Park Square, Milton Park, Abingdon, Oxon, OX14 4RN

CRC Press is an imprint of Taylor & Francis Group, LLC

ISBN: 9781041283171 (hbk)
ISBN: 9781041280781 (pbk)
ISBN: 9781003758822 (ebk)

DOI: 10.1201/9781003758822

Typeset in Times
by codeMantra

Contents

About the Authors

E. Scott Geller is an Alumni Distinguished Professor Emeritus at Virginia Tech, U.S. Over 56 years, he taught and conducted research in the Department of Psychology, directing the Center for Applied Behavior Systems. Dr. Geller has authored or coauthored over 50 books and 700 publications focused on behavior-change interventions to enhance quality of life. He is a Fellow of multiple psychological associations and has received numerous awards, including Virginia's Outstanding Faculty Award and lifetime achievement honors.

Krista S. Geller is Safety Director at Kiewit, NE. With a lifelong connection to applied behavioral science, she earned her Master's and Doctorate degrees in Human Development at Virginia Tech, focusing on family studies, gerontology, and pet–people relationships. Dr. Geller has worked globally as a People-Based Safety and Human Performance Improvement consultant, inspiring employees to actively care for safety. She is the President of GellerAC4P, Inc. and has authored research on behavioral safety and wellbeing. At Kiewit, she develops strategies to enhance safety, communication, and leadership through behavior analysis and interpersonal feedback.

Foreword

Dr Andrew Sharman

Across my career working with leaders and organizations around the world, one truth has remained constant: **people**—their beliefs, emotions, perceptions, motivations, and relationships—determine whether a workplace becomes safer, stronger, and more connected. Systems and procedures create structure, yes, but it is the *human dynamics beneath the surface* that shape the daily reality of safety.

This is a theme I've explored before in my own work on the social, emotional, and cultural forces that influence risk and behavior. In my writing and research, especially in my book *Naked Safety: Exploring the Dynamics of Safety in a Fast-Changing World*, I've argued that safety is not merely a technical or regulatory undertaking—it is fundamentally a human one. It depends on how people think, how they feel, how they make meaning of their environment, and how they relate to one another.

Few people have contributed more to this understanding than Dr E. Scott Geller. For more than 50 years, Scott has devoted his scholarship and his heart to illuminating the psychology of human behavior. I've had the privilege of collaborating with him—most recently in our book, *Who Cares Wins: The Psychological Science of Transformational Leadership*, where we made the case that **the future of leadership must be more human**. We argued that leaders who show empathy, build connection, and create psychological safety unlock extraordinary performance and engagement.

The book you now hold continues this essential conversation—but with a distinct and powerful focus. Coauthored by Scott and his daughter Krista, *The Human Dynamics of Achieving an Injury-Free Workplace* explores the psychological forces that shape safe and unsafe behavior, and the leadership actions that transform those insights into meaningful improvements in culture and performance.

Despite being presented as a series of directives and objectives, make no mistake, this is not a manual of rules or procedures. It is a guide to understanding people.

Scott and Krista explain—with clarity, evidence, and compassion—the behavioral science behind why people act the way they do. They explore competence, attitudes, social norms, emotion, interpersonal influence, self-talk, and the subtle cues that leaders send, often without realizing it. These psychological dynamics determine whether employees speak up or stay silent, whether they feel confident or hesitant, and whether they engage proactively in safe behavior or merely comply.

One of the most powerful messages in this book is its reminder that leaders shape culture not only through their decisions but also through their *presence*: their tone, their reactions, their emotional self-management, and the everyday micro-moments that convey what really matters. This mirrors much of what I explored in *Naked Safety*, where I argued that safety is created not by paperwork but by people—their assumptions, their stories, their social interactions, and the unwritten norms that govern behavior.

In that sense, this book beautifully complements the broader body of work on human dynamics in organizational life. Where *Naked Safety* looked at the wider context—wellbeing, engagement, governance, leadership, globalization, and social responsibility—this book zooms in on the behavioral levers that leaders can influence directly every day. The two perspectives reinforce each other: one revealing the landscape, and the other providing the tools for navigating it.

Most importantly, Scott and Krista remind us that leadership in safety is ultimately an act of humanity. Leadership is not only about setting direction or enforcing standards. It is about understanding oneself—the inner narratives, habits, emotions, and assumptions that silently influence our impact—and then understanding others with equal care. Only then can leaders create the social conditions in which people feel respected, valued, competent, and confident.

A culture of safety emerges when leaders connect the science of behavior with the art of being human.

My wish for you, the reader, is simple:

Let this book challenge you.

Let it broaden your understanding of human behavior.

Let it strengthen your awareness of the interpersonal dynamics that shape culture.

And let it inspire you to lead with clarity, curiosity, and humanity.

Because when leaders truly understand the human dynamics of safety—when they build environments in which people feel confident, connected, and supported, and when they actively care for their people—everyone wins:

People feel safer.

Teams grow stronger.

Mistakes and injuries decline.

Performance improves.

And the organization moves forward with greater trust, courage, and collaboration.

This is the power of psychological science translated into actions leaders can perform every day.

Scott and Krista have produced a compelling and practical guide for anyone committed to building safer, more human workplaces—and I'm very pleased to commend it to you now.

Dr. Andrew Sharman
CEO and Chairman of the Board
International Institute of Leadership & Safety Culture (IILSC)

A Note from the Authors

Our vision: A work culture in which all employees—from the plant manager and work supervisors to the frontline workers—are self-motivated and empowered to actively care for the safety, health, and psychological wellbeing of themselves and their coworkers each and every day.

This workbook shows you how to activate, cultivate, and sustain an actively-caring-for-people (AC4P) work culture. How, by teaching you how to customize evidence-based principles of psychological science for promoting, motivating, and supporting total employee engagement for occupational health and safety (OHS).

For a combined 85 years, E. Scott Geller, Ph.D., Alumni Distinguished Professor at Virginia Tech, and his daughter, Krista S. Geller, Ph.D., Safety Director for Kiewit Construction, helped organizations achieve and sustain an injury-free workplace by applying evidenced-based principles of psychological science to optimize interpersonal human dynamics.

Introduction and Acknowledgments

For more than five decades, I (Scott Geller) have taught the evidence-based principles and applications of psychological science (or humanistic behaviorism) at Virginia Tech, and in workshops and keynote addresses at regional and national safety conferences; and I have documented the research-based lessons in numerous books, manuals, book chapters, magazine columns, and professional research journals. While those presentations might have been effective at teaching the human dynamics of improving occupational health and safety (OHS) at some level, two limitations have reduced large-scale ownership and implementation of the research-based information delivered.

First, many presentations—especially in books, book chapters, and journal articles—have been too academic and detailed to hold the attention of readers looking for practical application strategies. Who has time to wade through a lengthy book that takes up more space describing the theory and research underlying a process than detailing relevant action plans? Moreover, the presentation of a particular behavior improvement technique does not always fit a reader's specific needs and/or the relevant context in which to apply evidenced-based principles from psychological science.

Attendees of my presentations at numerous safety conferences have expressed these limitations with such statements as "I appreciate the theory and principles presented by the professor, but I don't know how to apply his teaching." or "I am impressed with the amount of information presented and the numerous possibilities, but I don't know what to do with all of this at my workplace. How can I get my people engaged in using that information?"

As a behavior-based safety (BBS) human performance culture change agent and leadership consultant for more than a decade, Krista has heard similar concerns. Numerous times she challenged me with the same basic feedback:

> Dad, participants claim to understand the basics of behavior-based or people-based safety, but they seem to lack the authentic leadership and empowerment to design and implement a practical plan for their own work culture. We need to create something people can use in the field for their unique safety concerns and issues.

DOI: 10.1201/9781003758822-1

This education/training workbook addresses this legitimate concern in the best way possible. How? By not only explaining the human dynamics of achieving and sustaining an injury-free workplace, but providing questions for interpersonal dialogue and practice exercises that enable participants to customize evidence-based principles for their jobsites. In other words, this workbook instructs and inspires you to develop and apply an effective plan for improving the human dynamics of OHS at your workplace.

We recognize and thank the many safety leaders who have raised the critical issue of practical application over the years and incited the preparation of this innovative workbook. These inspirational individuals include the staff of Safety Performance Solutions (SPS) who have been teaching the BBS and actively caring for people (AC4P) principles and applications covered in this workbook to organizations worldwide since 1994. Their queries and insights from the field not only indicated the need for this workbook but also influenced its contents. Indeed, SPS has verified the social validity of the contents of this workbook by using the evidence-based principles to reduce the injury rates of numerous companies worldwide for more than two decades.

Moreover, we are extremely grateful for the dedication of the Virginia Tech research students and associates in the Department of Psychology, Center for Applied Behavior Systems (CABS). Since 1987, these individuals have collected and analyzed endless streams of field data to test the impact of various AC4P interventions, and thereby informed the design of more effective techniques to increase the frequency and improve the quality of AC4P behavior.

The support system of CABS served as a "think tank" for considering innovative approaches to understand and influence the human dynamics of interpersonal compassion and to develop research procedures for increasing AC4P behavior in various settings. I (Scott) am particularly beholden to my most recent and current graduate students: Samuel Browning, Nick Flannery, Trevin Glasgow, Zack Mastrich, and Jack Wardale; our research scientist: Dr. Erica Feuerbacher; and our more recent CABS coordinators: Caroline DeDecker, Ellie Townsend, Chris Esquivel, Loralee Hoffer, Jordan Oliver, Tyler Parker-Rollins, and Emily Warren.

I am indebted to the long-term advice, alliance, and friendship of Dave Johnson who edited this workbook to optimize readership. Dave was the editor of my first safety book—*The Psychology of Safety*—two subsequent volumes on people-based safety, and my textbook on applied psychology (Geller, 1996, 2005, 2008, 2016). Moreover, we coauthored a volume that teaches relevant humanistic behaviorism for healthcare workers—*People-based patient safety: Enriching your culture to prevent medical error* (Geller & Johnson, 2008).

Many others have inspired the development of this workbook, verifying the need for a teaching/learning process that enables participants to customize BBS and AC4P principles for their culture, and providing inspiration and guidance. Among these supportive colleagues, a few stand out: John Drebinger, Rodney Garringer, Nancy Kondas, Tim Ludwig, Charlie Morecraft, Denise Murray, Steve Roberts, and the late Bob Veazie.

Several special people have substantially influenced Krista's years in the business world, but one in particular kicked off Krista's safety career—the late Linda Berryman. She taught Krista what it means to have the courage to "stop a job" when safety is compromised, how to drive results, and how to maintain open communication with your team.

We are also grateful for the creative and instructive illustrations by George Wills of Blacksburg, VA. Since 1990, Scott Geller's teaching, textbooks, and workbooks have benefited from the artistic talents of George Wills.

Thank you all very much, including our many supportive colleagues and friends whose names we did not list. The synergy from your advice and inspiration empowers a laudable legacy: BBS and AC4P principles and practices readers can use to enrich their lives and contribute to cultivating a work culture where AC4P behavior is the social norm and an injury-free workplace is expected and commonplace.

OVERVIEW

An injury-free workplace requires effective ongoing engagement of all employees actively caring for the safety of themselves and their coworkers. This research-based OHS workbook explains and illustrates how to make that happen. Readers will learn ten safety-outcome objectives, each containing six evidence-based safety action directives relevant to successfully address the human dynamics of injury prevention and cultivate an AC4P culture.

The 60 safety action directives elucidated in this workbook reflect a variety of domains from psychological science, including the behavioral science principles of positive versus negative reinforcement, observational learning, and behavior-based feedback. Other lessons reflect humanism, including empathy, emotional intelligence, interdependency, systems thinking, psychological safety and emotional intelligence, and self-transcendence. We derived additional safety directives from social psychology, including principles of social influence, and the dynamics of group decision-making.

The ten teaching/learning safety objectives selected for this workbook are directly relevant for improving the human dynamics of OHS, and each objective reflects a distinct teaching/learning outcome—an OHS improvement proficiency relevant for cultivating an AC4P work culture by applying six safety action directives per objective.

Each safety objective and relevant directives include explanatory text and elucidating illustrations, a conclusion, discussion questions designed to inspire critical thinking and applications of specific safety directives, and relevant references from the research literature. We present each safety objective and six pertinent safety action directives as separate chapters in this workbook. A workshop instructor or facilitator could elaborate on the contents of this workbook, facilitate relevant discussion, and answer questions from participants.

The first safety objective and six directives provide an overview of the complexity of human nature. Consider this chapter an overview of the fundamental challenge addressed in this workbook—how to optimize individual and interpersonal human dynamics for OHS.

Safety Objective 2 with six directives initiates our focus on behavior by addressing the impact of behavioral consequences on occurrences of at-risk and safe behavior, and by illustrating how to apply consequences effectively to motivate occurrences of injury-preventive behavior.

The six safety directives for Safety Objective 3 connect principles of humanistic psychology to BBS in order to address the critical human dynamics of perceived empowerment and self-motivation. We provide answers to two critical questions: (a) What does it mean to feel empowered and self-motivated? and (b) How can people promote and support safety-related empowerment and self-motivation within themselves and among others?

The next safety objective and six safety directives are applicable to conduct a more accurate, unbiased, and useful analysis of a close call, injury, or property damage. Then, Safety Objective 5 and relevant safety directives address the potential application of six social influence principles to enhance employee engagement for OHS.

The six action directives for Safety Objective 6 address the absolutely crucial interpersonal coaching component of improving behavior, which many organizations do not accomplish effectively when attempting to implement a BBS process. The next objective and six directives teach the value of practicing interdependency and systems thinking to achieve synergy-an optimal OHS system with the whole greater than the sum of its parts.

Safety Objective 8 introduces the academic concept of humanistic behaviorism. Humanistic behaviorism defines the research-based foundation of the psychological science reflected throughout this workbook. Note the connection to actively (*behaviorism*) caring (*humanism*) for people (AC4P).

Safety Objective 9 addresses the cultivation of a healthy OHS mindset within oneself and among others. The safety directives for this objective explain why and how to (a) accept *stress* but stifle *distress*; (b) distinguish between personal *control* and personal *concern*; (c) accept Type A *behavior* but stifle Type A *emotion*; (d) apply intrapersonal conversation and constructive self-talk; and (e) adopt a lifelong mindset of continuous learning and teaching.

The final safety objective summarizes each of the prior safety objectives and most of the directives by explaining and illustrating their relevance to cultivating and sustaining a total safety culture (TSC). Here we explain three basic and practical strategies for initiating, achieving, and nurturing a TSC.

Discussion questions are included at the end of each chapter (i.e., a safety-outcome objective) to inspire reader engagement and bring the safety/life lessons to life. It would be very beneficial to discuss answers to these teaching/learning questions among members of a group of workbook readers or workshop participants with support from a facilitator with some knowledge of psychological science. However, individual readers can certainly write answers to the various questions for significant benefit to their personal learning and real-world application of the safety action directives per each chapter, resulting in a safety teaching objective and a valuable learning outcome.

Please note that each safety objective or chapter is essentially independent, and can be understood and applied without the information conveyed in any other chapter. Unlike most textbooks, the chapters in this workbook are not sequenced so the information in one chapter builds from the content of earlier chapters. Therefore, you could begin reading this workbook at any chapter, depending upon your current interest or a desired safety outcome (or objective). Likewise, a workshop facilitator could select a few of the chapters to cover for a particular safety topic. For example, a workshop on BBS might include three chapters—Safety Objectives 2, 6, and 8. A workshop on incident analysis would focus on Safety Objective 4, and a group discussion of "safety culture" would only need to cover the contents of the last chapter in this workbook—Safety Objective 10.

We estimate it would take at least three eight-hour workshop days to cover adequately all ten safety objectives presented here. Thus, most applications of this workbook will likely involve a select number of safety objectives. Of course, a ten-week series of three to four-hour OHS workshop discussions per week would be an ideal approach to covering the entire contents of this workbook. Although the order of presenting the safety objectives can vary, as indicated earlier, we recommend starting with the first chapter because it introduces the complexity of human nature and hence the unique challenges involved in improving the human dynamics of OHS and achieving an injury-free workplace. We also suggest covering Chapter 10 last, because this Safety Objective on cultivating a TSC reviews much of the content presented in earlier chapters.

THE LANGUAGE OF HUMANISTIC BEHAVIORISM

Some language of psychologists is unique and quite different from that used in everyday discourse. Moreover, behavioral scientists commonly use terminology that is uncommon in other domains of psychological science. Plus, we have coined certain AC4P terms and defined some related psychological concepts in distinct ways. Thus, immediately before the Index, we provide a Glossary of Key Terms used throughout this OHS workbook.

Even if you have had one or more courses in psychology, it could be useful to review the terms and definitions in this Glossary. We explain many terms with a behavioral and/or an AC4P focus, and this often results in a distinct difference from the more common usage of a term. For example, operational (i.e., behavioral or observable) definitions are offered whenever relevant and appropriate.

You might benefit from referring back to this Glossary of Key Terms when coming across particular behavioral science or AC4P jargon in the description of a safety objective or the application of a safety directive. In fact, it is useful to have a working definition of relevant key terms when discussing answers to the questions designed to activate critical thinking, interpersonal conversation, and/or action planning relevant to a particular safety objective and/or directive.

A Glossary term is bold and italicized in this text when we use it for the first time after this overview. These key terms are also bold in the Index. For emphasis later in the text, a Glossary term, as well as other terms, might be italicized without being bold. We use quotations to reflect words spoken by an individual or to call attention to unique or instructive language.

REFERENCES

Geller, E. S. (1996). *The psychology of safety: How to improve behaviors and attitudes on the job*. Radnor, PA: Chilton Book Company.

Geller, E. S. (2005). *People-based safety: The source.* Virginia Beach, VA: Coastal Training and Technologies Corporation.

Geller, E. S. (2008). *Leading people-based safety: Enriching your culture.* Virginia Beach, VA: Coastal Training and Technologies Corporation.

Geller, E. S. (Ed.). (2016). *Applied psychology: Actively caring for people.* New York: Cambridge University Press.

Geller, E. S., & Johnson, D. (2008). *People-based patient safety: Enriching your culture to prevent medical error.* Virginia Beach, VA: Coastal Training and Technologies Corporation.

1 Safety Objective 1
Appreciate the Complexity of Human Nature

Safety Directive 1: Start with behavior.
Safety Directive 2: Consider attitude for long-term behavior improvement.
Safety Directive 3: Distinguish between sensation and perception.
Safety Directive 4: Use imagery to motivate the occurrence of safe behavior.
Safety Directive 5: Apply cognition or self-talk to support safe behavior.
Safety Directive 6: Support the interdependent dynamics of occupational health and safety.

No one can deny the critical role of human dynamics when analyzing contributing factors to a workplace injury and when developing a safety **intervention** to prevent injuries and improve occupational health and safety (OHS). However, many safety consultants and professionals seem to believe that addressing the human side of safety involves only "culture/climate change," "mindset modification," "attitude adjustment," "increased enforcement," "self-accountability," or "behavior-based observation and feedback."

While each of these safety intervention themes is certainly relevant to improving the human dynamics of safety, taken alone each is limited and insufficient. One might ask, "Which of those should be included in a mission to improve the safety culture of a workplace?"

A narrow perspective regarding the psychology of injury analysis and prevention is certainly understandable, given the extensive marketing of select tools or intervention approaches delivered by safety consultants and trainers with limited **education** and experience in psychological science.

We are not criticizing nor demeaning those diverse approaches to address the human dynamics of safety. Indeed, many of those approaches are quite essential for cultivating a safety-engaged workforce. But many safety intervention approaches implicate a rather narrow view of the human side of safety. This first safety objective introduces seven dimensions of people, each reflecting intervention potential for OHS.

DOI: 10.1201/9781003758822-2

THE BASIC ID ACRONYM

After Scott taught his introductory psychology students the various approaches to psychotherapy, from Sigmund Freud's psychoanalysis to B.F. Skinner's behavior analysis and Aaron Beck's cognitive therapy; he introduced BASIC ID—an acronym reflecting the complexity of human nature. Scott pointed out that few clinical science approaches to counseling or psychotherapy address each of these human dynamics.

Of course, the same is true for handling the human side of safety. Although safety professionals do not delve deeply into the analysis and treatment of psychological issues, considering the following seven dimensions of human nature could influence a more comprehensive and successful approach to cultivating an actively caring for people (AC4P) work culture.

"B" for Behavior

Obviously, workers' on-the-job behavior influences the probability of an injury, and workers' AC4P behavior enhances the success of any injury prevention process. Thus, behavior-based safety (BBS)—especially peer-to-peer observation and feedback—has become a worldwide phenomenon that successfully reduces workplace injuries by engaging employees in interpersonal, behavior-focused coaching (Geller, 2024; Sulzer-Azaroff & Austin, 2000).

As we explain in this workbook, many BBS programs have not been nearly as successful as they could have been. Why? They did not consider dynamics of human nature beyond overt or observable behavior. In a BBS process, for example, workers observe each other's safety-related behavior and record their observations on a customized checklist. Later, someone enters that information on a computer file for analysis that includes comparisons of safe-behavior percentages between work teams and across facilities. However, the most critical component of a BBS observation-and-feedback process is often omitted—interpersonal behavioral coaching from observer to worker.

Plus, when peer-to-peer coaching does occur, delivery of the behavioral **feedback** is often inadequate, especially when providing corrective feedback for observed at-risk behavior. This workbook explains how to give **supportive** and **corrective feedback** effectively—so feedback is accepted, appreciated, and applied (Safety Objective 6).

"A" for Attitude

Considering the prior example of BBS coaching, the delivery and acceptance of behavioral feedback requires the right attitude from everyone involved.

A BBS program should not be viewed as a top-down requirement, but rather as an employee-owned process in which the participants feel empowered and self-motivated (Safety Objective 3). Indeed, the success of any intervention process implemented to improve OHS depends on an interdependent (Safety Objective 7), hopeful, and constructive attitude or mindset (Safety Objective 9).

Please note that an attitude is much more than an opinion. It is a deep-seeded personal perspective quite resistant to change. Consider that an attitude influences affect, behavior, and cognition—the ABCs of attitude. First, an attitude connects to a person's affect or emotion. People feel closer to those who share their attitude about something and might feel a touch of anger, frustration, or sadness toward those who hold an opposite attitude. And naturally, our attitude about something influences concomitant behavior and cognition (i.e., thinking or **self-talk**).

"S" for Sensation

Obviously, to achieve and sustain an injury-free workplace, workers often need to strategically apply their five senses: seeing, hearing, feeling, and sometimes even smelling and tasting. For example, distracted driving—all too common a cause of vehicle crashes—occurs when drivers neglect the appropriate application of their visual sensations. "Their eyes were not consistently on task," as the safety professional would say.

Competent and safe completion of almost any job requires visual acuity, hand-eye coordination, and often a keen sense of timing (e.g., when adjusting machinery). Sometimes, it is necessary to react quickly and accurately to a dangerous situation in order to prevent an injury. Alert sensation has prevented many mishaps.

Selective sensation. We do not see and hear all of the stimuli in our immediate surroundings. We ignore irrelevant sights and sounds, and focus our attention on the stimuli relevant to the task on hand. When individuals get distracted from task-relevant sensations (e.g., when driving), they put themselves at risk for incompetent behavior and an injury. Thus, safety professionals urge workers to tune in all relevant sensations and tune out task-irrelevant sensations. Safe and competent work requires selective sensation. Obviously, a comprehensive analysis of an injury should consider the potential role of sensory distractions or inattention to task-relevant stimuli.

Selective perception. Most readers have probably heard the term "selective perception," and perhaps have used this concept to explain a misunderstanding or misperception. Please note that the term "selective perception" is actually redundant because all perception is selective—biased by personal experience, as well as by both dispositional (nature) and situational (nurture) factors. Simply put, perception is an individual's *interpretation* of a sensation experienced through any or all of the five senses. Hence, the sensation/perception course offered by the psychology department of many colleges and universities covers both the physiology of our sensory system (i.e., sensation) and the factors that influence our interpretation of sensation (i.e., perception).

Figure 1.1 illustrates "selective" perception and reflects the concept of premature cognitive **commitment**, introduced by Ellen Langer (1989). Specifically, the perceptual bias (or biased perception) of the male driver has made him unaware of the road hazard ahead of him, as well as the attempt of the female driver to warn him. Why is this premature cognitive commitment?

FIGURE 1.1 Perceptual bias can be risky.

First, it is premature—occurring before adequate diagnosis, analysis, and decision-making. Second, it is cognitive—a thinking process that influences perception, attitude, and behavior. Finally, it can be a commitment—a relatively permanent perspective or mindset that filters the information received, biases interpretation of sensation, and shapes behavior. Some refer to such perception as "confirmation bias"—typified by the conflicting verbal behavior of Republicans versus Democrats and vice versa.

"I" FOR IMAGERY

Imagery is using our "mind's eye" to picture situations without actually being there. We use mental imagery every day. When we look forward to a particular event, we use imagery. Sometimes we visualize the expected outcome of an upcoming event, and that imagery affects our motivation. Picturing a pleasant **consequence** can lead to excitement, even an "emotional high"; but imagining negative outcomes can elicit fear and motivate avoidance behavior.

Before performing, athletes practice their sport mentally; actors run through their lines and stage positions in their mind's eye; surgeons mentally rehearse the steps of a complex operation; and musicians imagine playing or singing the right notes on key and on time. Many public speakers visualize themselves delivering their lines proficiently just prior to their actual presentation.

Research has demonstrated significant benefits of mental rehearsal (e.g., Cumming & Williams, 2012; Wakefield & Smith, 2012), whether practicing an athletic skill, an occupational task, or a script of verbal dialogue. It is not clear whether the mental rehearsal actually strengthens the correct behavior or merely increases one's motivation to perform at a higher level. We don't know why mental rehearsal improves **performance**, only that it does.

The more vividly individuals can imagine themselves performing desired behaviors, the greater the beneficial impact of this technique on actual performance. Now what does that mean for occupational safety?

Imagery and safety. The authors could not find any research on the effects of imagery on safety-related behavior. But given the variety of behaviors shown to benefit from imagery, it seems obvious employees can use imagery to prevent an injury. People use it to anticipate and prepare for upcoming events. We can use imagery to direct our behavior (as an **activator**) or to motivate our behavior (as a consequence). More specifically, for safety self-management, mental imagery can be used to:

- clarify safety objectives
- enhance **self-motivation** to choose the safest behavior
- build **self-efficacy**, **personal control**, and/or **optimism**
- rehearse safe acts and AC4P behavior
- **reward** oneself for effective self-management

Imagery can activate a chain of safe operating procedures (SOPs), as well as motivate action. The motivation comes from imagining potential consequences following safe versus at-risk behavior. Figure 1.2 depicts a negative consequence one might visualize in order to motivate the use of personal protective equipment (PPE). The image also suggests the particular safe behavior needed to avoid a particular injury. With this imagery, the worker motivates himself to use the necessary PPE. Thus, the image is both directing and motivating.

It is important to be active in your image. Don't see yourself as a passive observer watching a movie starring you. Rather, imagine yourself acting within the complete **activator-behavior-consequence (ABC)** framework.

That is, see the activators in the situation that cue the desired behavior; visualize yourself actually performing the safe act(s); then imagine positive feeling states (i.e., the rewarding consequence) from setting the safe example for others and acknowledging safety as a value.

Before imaging the ABC sequence, you might find it motivating to visualize the negative consequence of getting hurt, as illustrated in Figure 1.2. When reaching for that skill saw, for example, imagine getting a finger caught in the blade. Imagine the ringing in your ears getting worse after not using hearing protection. Similarly, imagining a vehicle crash can inspire drivers to remind all passengers in their vehicle to "buckle up."

FIGURE 1.2 Imagined consequences can motivate behavior.

Sharing an image. It can be useful to share personal experiences in ways that conjure up a motivating image. In fact, personal testimonies of injuries or close calls are powerful motivators because the listeners can get a mental image of the incident. They can readily visualize the speaker in the precarious situation described, especially if the presenter gives a passionate and realistic delivery. Then, listeners can put themselves or a family member in the situation that caused negative consequences for the speaker. Of course, it is essential to focus on the specific safety-related behaviors that would have prevented the injury described if the presenter had performed them.

Consider sharing the use of mental imagery for injury prevention, as reviewed here, and encouraging personal testimonials of injuries or close calls that could have been prevented by certain action—the avoidance of a particular at-risk work practice and/or the performance of a desirable safe behavior. You can use mental imagery to direct and motivate safe behavior in yourself and others. However, an image is more directing and motivating when words are associated with the image. This is covert behavior or cognition—the next human dynamic of BASIC ID.

"C" for Cognition

Cognition implies thinking or covert self-talk. Cognition associated with the image of a devastating injury can direct overt safety-related action. "To eliminate my image of a worker falling from that roof, I must promote and support the use of appropriate fall protection."

Cognitive therapists attack the irrational or negative thinking of their clients in order to "think people into relevant behavior change" (Beck, 1976, 1993). From the same framework, safety leaders attempt to (a) prevent cognitions that can provoke at-risk behavior, (b) enhance safety-related thinking that can activate injury-prevention behavior, and (c) incite self-talk that can inspire occurrences of AC4P behavior on behalf of another person's safety.

In addition to personal safety, cognitions influence and reflect **self-esteem**. We can focus our self-talk on the good things people say about us or on other people's critical statements about us. Our self-talk reflects our perceptions and can increase or decrease how we feel about ourselves (Tod, Hardy, & Oliver, 2011). As illustrated in Figure 1.3, negative self-talk can interfere with the kind of positive experiences that should incite positive intrapersonal communication and increase self-esteem.

FIGURE 1.3 Negative self-talk can ruin a good thing.

"I" FOR INTERPERSONAL

An AC4P safety culture requires ongoing interpersonal prosocial support for safety—from identifying and removing injury-related hazards to interpersonal AC4P mentoring, coaching, and inspiration for injury prevention. Researchers have demonstrated convincingly that interpersonal support enhances personal health, happiness, and life satisfaction (Cohen & Willis, 1985; Isen, 1987; Young & Glasgow, 1998). Happy workers are more productive and safer.

As illustrated in Figure 1.4, a win/lose mindset is often more popular than a perception of win/win. However, an **AC4P culture** requires a win/win interdependent mindset. Workers need to depend on each other to maintain an injury-free workplace. This includes peer-to-peer BBS coaching (Safety Objective 6). While you depend on team members to complete their task assignments safely, the others depend on you to do the same.

FIGURE 1.4 Our culture often promotes win/lose.

When it comes to OHS, some workers are reluctant to offer another person a safety reminder and to receive advice for their own safety. Why, because they consider safety a matter of individual or personal responsibility. This perception is reflected in such self-talk as "If Mary and Bill want to put themselves at risk, that's their problem, not mine."

Thus, some individuals need a change in perception—a paradigm shift—to facilitate and support the occurrence of interdependent safety coaching (Safety Objective 6). Everyone needs to consider safety coaching as a shared responsibility—an interdependent opportunity to prevent injuries throughout their work culture. This calls for a perceptual shift from win/lose individualism to win/win collectivism.

"D" for Drugs

When Scott asks his students to guess the human dimension reflected by the letter "D," someone inevitably yells out "drugs," presumably as a joke. Scott's reply—"Absolutely right, but I'm not talking about those illegal drugs you sniff or inject. How about those over-the-counter drugs—alcohol, nicotine, caffeine, or prescription drugs for pain—that influence each of the prior human dynamics we've reviewed?"

Scott urges his students to pay attention to the physical and psychological impact of those common drugs. Of course, this is not enlightening information for most safety professionals, given that many have dealt with drug issues among employees, resulting in occasional referrals to an Employee Assistance Program.

Conclusion

As emphasized above, we did not review the complexity of human nature and the difficulty in changing people and their culture to challenge any particular intervention approach. Rather, our purpose was to prompt consideration that improving the human dynamics of OHS is more difficult than often marketed. In fact, most safety professionals, consultants, and trainers have not been educated nor trained sufficiently to intervene on behalf of the seven human dynamics reviewed here.

After a combined 85 years in researching and teaching psychological science, we recommend the following. Start with behavior by implementing a BBS observation-and-feedback coaching process, but recognize the need to solicit and sustain employee engagement through their supportive attitudes, perceptions, cognitions, and interpersonal AC4P behavior.

Your mission: "Act people into beneficial safety-related attitudes, perceptions, cognitions, images, and interpersonal support." Work with consultants and trainers who believe in this mission and want to help make it happen. The remaining 9 safety objectives and 54 safety directives teach you how to make that happen, with hope you will share these research-based principles and intervention techniques with others.

DISCUSSION QUESTIONS

1. What safety-relevant concepts or intervention approaches related to human dynamics or psychological science had you already learned before receiving this workbook?

2. What have you already learned about BBS?

3. Do you believe BBS is effective at preventing injuries? Why or why not?

4. Discuss the difference between sensation and perception with a real-life example.

5. How could you combine imagery and cognition to prevent a personal injury to yourself?

6. How have you used *intra*personal conversation or self-talk to influence your safety-related behavior and/or the safety of others?

7. What does "*inter*personal" mean with regard to OHS?

8. What are some barriers to cultivating an interpersonal AC4P safety culture at your work facility?

9. Would you recommend intervening to “act people into safe thinking” or to “think people into safe acting”? Please explain your selection.

10. In what ways do the safety-related signs or slogans used in your work facility to promote OHS connect to one or more of the seven human dimensions represented by the BASIC ID acronym?

REFERENCES

Beck, A. T. (1993). Cognitive therapy: Past, present, and future. *Journal of Consulting and Clinical Psychology, 61*(2), 194–198.

Beck, A. T. (1976). *Cognitive therapy and the emotional disorders.* New York: International Universities Press.

Cohen, S., & Wills, T. A. (1985). Stress, social support, and the buffering hypothesis. *Psychological Bulletin, 98,* 310–357.

Cumming, J., & Williams, S. E. (2012). Imagery: The role of imagery in performance. In S. Murphy (Ed.), *Handbook of sport and performance psychology* (pp. 213–232). New York: Oxford University Press.

Geller, E. S. (2024). *Actively caring for safety: The psychological science of injury prevention.* Boca Raton, FL: CRC Press.

Isen, A. M. (1987). Positive affect, cognitive processes and social behavior. *Advances in Experimental Social Psychology, 20,* 203–253.

Langer, E. J. (1989). *Mindfulness*. Reading, MA: Addison-Wesley.

Lazarus, A. A. (1971). *Behavior therapy and beyond.* New York: McGraw-Hill.

Lazarus, A. A. (1976). *Multimodal behavior therapy.* New York: Springer.

Sulzer-Azaroff, B., & Austin, J. (2000). Does BBS work? Behavior based safety and injury reduction: A survey of the evidence. *Professional Safety, 45,* 19–24.

Tod, D., Hardy, J., & Oliver, E. (2011). Effects of self-talk: A systematic review. *Journal of Sport & Exercise Psychology, 33,* 666–687.

Wakefield, C., & Smith, D. (2012). Perfecting practice: Applying the PETTLEP model of motor imagery. *Journal of Sport Psychology in Action, 3*(1), 1–11.

Young, F. W., & Glasgow, N. (1998). Voluntary social participation and health. *Research on Aging, 20,* 339–362.

2 Safety Objective 2
Apply Consequences to Motivate OSH Behavior

Safety Directive 7: Use more positive than negative consequences.
Safety Directive 8: Apply "if-then" and "now-that" rewards.
Safety Directive 9: Distinguish between extrinsic and intrinsic consequences.
Safety Directive 10: Focus on intrinsic consequences whenever possible.
Safety Directive 11: Appreciate the impact of emotional intelligence (EQ).
Safety Directive 12: Consider the relevance of the AC4P Movement for OHS.

Every reader realizes the significant impact of human behavior on both the occurrence and the prevention of workplace injuries. Almost every safety professional has heard the term **BBS**. However, how many readers understand and appreciate the foundation of BBS, beyond the systematic observation and documentation of at-risk and safe work behavior, and the delivery of relevant behavioral feedback?

Safety Objective 2 addresses a fundamental principle of **applied behavioral science (ABS)** and the research-based foundation of BBS—the consequences of behavior. Behavioral consequences are **extrinsic** or **intrinsic**, and they either help or hinder the prevention of workplace injuries.

MOTIVATION BY CONSEQUENCES

B.F. Skinner, the founder of the research and scholarship domain of ABS, communicated and validated this simple but profound principle—**selection by consequences** (Skinner, 1938, 1974). Indeed, Skinner and his followers have demonstrated consistently that behavior occurs to gain a positive consequence and to avoid or escape a negative consequence. Plus, behavior can be modified by changing the consequence that follows it.

Three critical aspects of behavioral consequences are relevant for motivating human behavior: (a) Positive consequences are more effective than negative consequences at improving behavior; (b) The best consequences activate and/or support self-motivation or **self-directed behavior**; and (c) Natural or intrinsic consequences have special advantages over extra or extrinsic

DOI: 10.1201/9781003758822-3

consequences, as we will explain with regard to motivating human behavior for injury prevention.

The ABC Model

ABS is founded on a basic ABC model. "A" represents an activator or the antecedent event(s) immediately preceding a target behavior—"B." "C" denotes the consequence—the event following the behavior that influences its recurrence. An activator is an **incentive** when it announces the availability of a positive consequence—a **reward**—following the occurrence of a particular desirable behavior. In contrast, an activator is a **disincentive** when it announces the availability of a negative consequence—a **penalty**—following the occurrence of a designated undesirable behavior.

The sign in Figure 2.1 is an activator and could be an incentive or a disincentive, depending on the perception of the driver. In this case, the driver perceives the sign as a disincentive because she expects other drivers to report only occurrences of undesirable or unsafe driving behaviors. Why doesn't the driver expect a phone call for her safe driving? Is this because the driver will not be driving safely and would not be eligible for a positive phone call? Of course not, but it is certainly disappointing that people are more likely to notice and criticize undesirable behavior than to notice and recognize desirable behavior.

FIGURE 2.1 Activate behavior with a consequence.

Our culture influences a focus on undesirable behavior and the use of negative consequences to eliminate such behavior. The news and social media report much more about unwanted than wanted behavior, and the government approach to controlling behavior is essentially to pass a law and enforce it. We live in a "Click it or Ticket" society. However, ABS has demonstrated various advantages of using positive over negative consequences to increase the quantity and/or improve the quality of behavior.

If-Then vs.Now-That Rewards

Both research and commonsense tell us it is extremely beneficial to call attention to behavior that we want to see occur more often and to recognize those desirable behaviors. Behavioral scientists refer to such positive consequences as **now-that** rewards as opposed to **if-then** rewards referred to above. An if-then reward reflects an incentive: "*If* you perform this behavior, *then* I will give you this reward."

A now-that reward is different. There is no incentive—just a reward or a sincere statement of appreciation for the desirable behavior observed. The delivery of now-that recognition is simple and straightforward. You simply specify the desirable behavior observed and offer sincere praise, recognition, and/or gratitude. The result: You might increase occurrences of the recognized behavior, and you will surely enhance the person's perception of personal competence and self-motivation. As Dale Carnegie (1936) emphasized in his classic best seller—"Everyone wants to feel important."

Extrinsic Criticism

Figure 2.2 reveals a criticism that some pop psychology books have used to discredit the use of positive consequences or rewards to increase the frequency of desirable behavior. Such negativity is reflected by the title of one of those books—*Punished by rewards: The trouble with gold stars, incentive plans, A's, praise, and other bribes* (Kohn, 1993). Authored by a journalist, this popular self-help book caused quite a stir among behavioral scientists, who realized that the benefits of using positive consequences to motivate occurrences of desirable behavior had been demonstrated consistently in numerous empirical studies and real-world applications since the 1950s.

Systematic reviews of the research related to that pop psychology critique of extrinsic rewards have revealed the following: (a) As discussed above, consequences desired by participants increase the frequency of the targeted behavior, and (b) When behavior-based rewards are removed, the target behavior decreases in frequency of occurrence, as commonsense indicates. However, contrary to the uninformed authors of books that demean the use of rewards,

FIGURE 2.2 Don't focus on extrinsic consequences.

when an extrinsic reward is discontinued, the target behavior rarely occurs less often than its pre-reward or baseline frequency (Cameron & Pierce, 1994; Eisenberger & Cameron, 1996).

Here is the most practical research-based conclusion regarding the use of extrinsic rewards to improve behavior: After using extrinsic rewards to get a desirable behavior started, the intervention agent should fade the positive consequence (i.e., give it intermittently after the target behavior). Emphasize the natural or intrinsic consequences, such as the meaningfulness or the long-term purpose of the target behavior.

The natural consequences of a target behavior should be identified at the start of a motivational process, but it often takes an extrinsic consequence to motivate initial participation and enable recognition of the natural longer-term benefits of the desirable behavior. The teacher in Figure 2.2 should certainly focus on the natural or intrinsic consequences of the situation—"Billy, how wonderful, you solved the math problem. You are becoming more and more competent at math; your parents will be so proud of you."

Regarding occupational safety, an extrinsic reward might be needed to motivate delivery of a thoughtful safety suggestion or reporting a close call or "near miss" (actually a "near hit"). However, consider the intrinsic motivating consequences of employees seeing an injury prevention change in the work setting or in the safety management system following their safety suggestion or the report of a close call.

THE POWER OF INTRINSIC CONSEQUENCES

Some natural (or intrinsic) consequences are immediate, while others are delayed. In Figure 2.3, each person sees immediate natural consequences—the fruits of their labor. However, mom and dad are likely also thinking about a delayed consequence, such as a large shade tree and a veggie-producing garden. While their son is experiencing an immediate intrinsic consequence—**positive reinforcement**—for his car-washing behavior, he might also be thinking of later impressing his date with a clean car.

FIGURE 2.3 Intrinsic consequences are naturally motivating.

Are you thinking that intrinsic or natural consequences are more powerful than extrinsic or extra consequences? Yes, your commonsense is correct because you have been there. You have experienced the flow of behavior motivated by immediate natural and positive consequences. Consider, for example, the entertainment from reading and watching TV, or the sense of accomplishment from playing a musical instrument, practicing a sport, preparing a meal, or applying artistic skills to drawing, painting, or woodworking behavior.

What about pushing those buttons on your computer, iPad, or cell phone? Some consider that behavior an addiction that causes distracted driving and even distracted walking. This discussion reveals a different reality. Such behavior is not a disease, but the outcome of devices that reinforce the flow of behavior with soon, certain, positive, intrinsic consequences. In addition to texting while driving and walking, intrinsic positive **reinforcers** support many other undesirable behaviors (e.g., excessive alcohol consumption, cigarette and marijuana smoking, and overeating). What about at-risk behavior?

Do more positive intrinsic consequences typically follow at-risk than safe behavior? A "Yes" answer to this question reflects Scott's early contribution to *Industrial Safety and Hygiene News*, entitled "Managing safety: Fighting human nature" (Geller, 1990). More comfort, convenience, and efficiency are three obvious intrinsic reasons for workers choosing at-risk over safe behavior. Realizing this, safety leaders attempt to increase the comfort, convenience, and efficiency of safe work behaviors.

Unfortunately, it is usually impossible for safe behavior to match the convenience and efficiency of at-risk behavior. Often there is a need to implement one or more safety-related interventions, from education/**training** programs to an **extrinsic motivator**—an incentive or a disincentive. However, education/training provides only an activator—a reminder. Even though positive consequences are more effective at changing behavior and improving attitude, negative consequences—with more apparent convenience and less financial expense—usually take precedence over incentive/reward interventions.

Fortunately, some individuals—particularly safety leaders—have the **emotional intelligence (EQ)** to avoid the enticement of soon, certain, and positive consequences for at-risk behavior.

THE POWER OF EQ

Figure 2.4 illustrates the difficulty in getting nonhuman animals to delay the immediate pleasure of experiencing an available positive consequence. Yes, dogs can be trained to wait awhile before devouring a morsel of desirable food. What about humans?

FIGURE 2.4 It is not easy to delay a positive consequence.

Have you heard a statement like “Your success in life depends upon your ability to delay immediate pleasures for larger rewards from long-term accomplishment”? Research indicates that one’s ability to keep working on a task in the face of temptation to engage in another activity with soon and certain positive consequences reflects EQ, and EQ is quite predictive of career success (Goleman, 1995, 1998).

To verify the direct relationship between one’s ability to delay immediate gratification for future success, Walter Mischel (2014) challenged four-year-old children with a “Marshmallow Test” to assess their impulse control. In a one-to-one context, a researcher handed a four-year-old a marshmallow and said s/he could eat it now or wait until later and receive two marshmallows. Some children ate the single marshmallow within a few seconds after the researcher left the room. The other children waited 15–20 minutes for the researcher to return with an extra marshmallow.

Most children who delayed immediate gratification for a delayed but larger reward did not just sit patiently and wait. Instead, they engaged in behaviors that apparently facilitated their self-discipline to resist the impulse for immediate pleasure. Some sang or talked to themselves; others played games with their hands and feet; and others covered their eyes or buried their head in their arms.

Fourteen years later, when these children were evaluated as adolescents during their last year of high school, those who had waited patiently at age four were far superior as students than were those who had failed the Marshmallow Test. They were more academically competent; they had better study habits; and they appeared more eager to learn.

They were better able to concentrate, to express their ideas, and to set goals and achieve them (Mischel, 2014). In fact, those higher achievers scored significantly higher on both the math and verbal portions of the scholastic assessment test (SAT)—by an average of 210 total points—than did the students who had not delayed gratification at age four (Shoda, Mischel, & Peake, 1990).

While it is tempting to consider EQ to be a function of nature, research indicates that EQ can be learned and nurtured through appropriate education/training that includes (a) techniques to encourage and support self-motivation and a perception of **empowerment** (Safety Objective 3), (b) relevant **behavior-based feedback** (Safety Objective 6), and (c) a context that facilitates an optimistic and success-seeking mindset (Safety Objective 9).

However, activating and motivating EQ for safety-related behavior is especially challenging. The safety professional cannot even promise a delayed reward for performing safety-related behavior. Safety leaders ask workers to do the right thing for OHS in order to avoid the *improbable* possibility of an injury. Indeed, the probability of an injury to an employee working at risk is extremely low, supporting the commonsense self-talk, "It won't happen to me." Such thinking is usually supported by the lack of an injury from performing relatively efficient, comfortable, and convenient at-risk behavior.

Consistently choosing safe over at-risk behavior requires a relatively high level of EQ. Here is a suggestion: Insert the mission of the AC4P Movement (www.ac4p.org) into discussions about substituting safe behavior for more intrinsically reinforcing at-risk behavior. Acknowledge the reality that the probability any one individual will be seriously injured while performing a particular at-risk behavior (e.g., texting while driving) is minuscule, supporting the intrapersonal conversation, "It won't happen to me."

However, ask the individual whether someone will eventually get hurt if many people in their work setting repeated the particular at-risk behavior many times. After the individual gives the obvious answer of "Yes," then ask, "Do you care?" (If the worker does not answer "Yes" to that question, you should seriously consider whether you want this person to continue working at your facility.)

When the person answers, "Yes, I care," you should suggest that the individual can *actively* care by always demonstrating safe behavior. You might refer to **observational learning** (addressed later in Safety Objectives 4 and 10). Plus, you could introduce the concept of a social norm (Safety Objective 10). Specifically, when the majority of the individuals in a culture perform a certain behavior, that behavior becomes the social norm. If an individual really cares about keeping all workers safe, s/he needs to actively care by always demonstrating the safe example for others to see.

Conclusion

One can appreciate the challenge of motivating occurrences of safe over at-risk behavior by considering the power of intrinsic or natural consequences. While natural or intrinsic consequences like relative comfort, convenience, and/or efficiency facilitate at-risk behavior, relative discomfort, inconvenience, and/or inefficiency can inhibit occurrences of the safe alternative. As a result, extrinsic incentive/reward or disincentive/penalty contingencies are often implemented, with negative consequences (e.g., a penalty for at-risk behavior) used more often than positive consequences to motivate the performance of safety-related behavior (e.g., if-then or now-that rewards).

Why are disincentive/penalty interventions more popular than more effective incentive/reward contingencies? The answer is similar to the rationale for the occurrence of at-risk behavior—convenience, efficiency, financial cost, and perhaps a lack of knowledge.

It takes EQ to (a) learn the research-based advantages of motivating with positive over negative consequences; (b) create innovative ways to implement a behavior-based intervention process for injury prevention; and (c) take the extra time and sometimes the extra financial expense to implement an extrinsic incentive/reward contingency to motivate participation in an injury prevention process (e.g., peer-to-peer observation and feedback; safety improvement suggestions and follow-up; and close-call reporting, analysis, and corrective action). This workbook was designed to activate and inspire a variety of positive and practical intervention considerations and applications.

DISCUSSION QUESTIONS

1. Why are negative consequences used more often than positive consequences to improve behavior at work, in school, at home, and throughout communities?

2. Which behavior change technique—the application of positive or negative consequences—will usually have the longest-term beneficial impact? Why?

3. Explain the difference between an "if-then" and a "now-that" reward with real-world examples?

4. Discuss the concept of "motivation" with examples of extrinsic and intrinsic consequences.

5. How have you seen extrinsic rewards used to promote OHS?

6. Why or why not were the extrinsic reward techniques identified for Question 5 effective?

7. Derive a plan for using extrinsic and intrinsic consequences to increase the frequency of a safety-related behavior.

8. Discuss EQ with regard to cultivating an AC4P culture to keep people safe.

REFERENCES

Cameron, J., & Pierce, W. D. (1994). Reinforcement, reward, and intrinsic motivation: A meta-analysis. *Review of Educational Research, 65*(3), 363–423.

Carnegie, D. (1936). *How to win friends and influence people* (1981 Edition). New York: Galahad Books.

Eisenberger, R., & Cameron, J. (1996). Detrimental effects of reward: Reality or myth? *American Psychologist, 51*(11), 1153–1166.

Geller, E. S. (1990). Managing safety: Fighting human nature. *Industrial Safety and Hygiene News, 24*(9), 14.

Goleman, D. (1995). *Emotional intelligence*. New York: Bantam Books.

Goleman, D. (1998). *Working with emotional intelligence*. New York: Bantam Books.

Kohn, A. (1993). *Punished by rewards: The trouble with gold stars, incentive plans, A's, praise, and other bribes*. Boston, MA: Houghton Mifflin.

Mischel, W. (2014). *The marshmallow test: Mastering self-control*. New York: Little, Brown.

Shoda, Y., Mischel, W., & Peake, P. K. (1990). Predicting adolescent cognitive and self-regulatory competition from preschool delay of gratification. *Developmental Psychology, 26*, 978–997.

Skinner, B. F. (1938). *The behavior of organisms: An experimental analysis*. New York: Appleton-Century.

Skinner, B. F. (1974). *About behaviorism*. New York: Alfred A. Knopf.

3 Safety Objective 3
Inspire Empowerment and Self-Motivation for OHS

Safety Directive 13: Distinguish between *managing* behavior and *leading* people.
Safety Directive 14: Employ the power of perceived choice.
Safety Directive 15: Enhance perceptions of personal competence.
Safety Directive 16: Promote and support interdependence.
Safety Directive 17: Help yourself and others feel empowered.
Safety Directive 18: Set empowering goals and apply consequences.

Achieving and sustaining an injury-free workplace requires relevant safety engagement from everyone. Workers need to be self-motivated and feel empowered to act continually on behalf of the safety and health of others. This includes proactive analysis of close calls and minor injuries (Safety Objective 4) and interpersonal behavior-based **AC4P coaching** (Safety Objective 6).

Of course, managers and supervisors need to be self-motivated and feel empowered, but they also need to facilitate self-motivation and perceptions of empowerment among employees. How can they do this? Psychological science offers evidence-based answers to that critical question. First, this safety objective addresses the issue of self-motivation and then explains the psychological definition of empowerment. Then, connections between self-motivation and empowerment are specified, including similarities and differences between these crucial human dynamics.

The research-based principles explained and illustrated here need to be discussed with colleagues in order to customize related applications for self-motivation and for leading others to achieve and maintain an AC4P culture—essential for achieving and maintaining an injury-free workplace.

LEADERSHIP AND SELF-MOTIVATION

Managing people is not the same as *leading* people. Still, both **management** and **leadership** are essential for cultivating an AC4P injury-free work

DOI: 10.1201/9781003758822-4

culture. Simply put, managers hold people accountable to perform desirable (e.g., safe) behavior and avoid performing undesirable (e.g., at-risk) behavior for OHS. They manage or motivate behavior with an external or extrinsic accountability intervention or system. In contrast, leaders inspire self-motivation by influencing particular person-states—perceptions, attitudes, and/or emotions—that facilitate self-motivation.

Self-motivation or self-direction often leads to **discretionary behavior**—more desirable behavior than requested or required. To be sure, safety managers can be safety leaders. While managers hold an assigned position that enables them to control certain motivating contingencies or behavioral consequences, anyone, regardless of their position, can be a leader. How, by promoting self-motivation in yourself and others (Geller, 2016). Psychological science provides evidence-based strategies to make that happen by increasing perceptions of **choice, competence, and community**.

Perceived Choice

At times, people need external activators (e.g., incentives or disincentives) and consequences (e.g., rewards or penalties) to keep them motivated. However, sometimes people develop self-motivation and self-accountability within the context of an extrinsic accountability system. That person-state includes the *perception* of choice, as reflected in Figure 3.1.

FIGURE 3.1 Choice is in the eyes of the beholder.

In fact, people have a need for autonomy, regardless of dispositional and situational factors (Deci, 1975; Deci & Flaste, 1995). **Participative management** means employees enjoy personal choice during the planning, execution, and/or evaluation of their jobs. In the workplace, managers often tell people what to do in order to be most efficient. It takes more time to involve people in the decision-making process and to promote perceptions of choice and inspire self-motivation.

Consider how language can influence a perception of external control or personal choice. Should managers "give mandates" or "set expectations"? Should they "demand compliance" or "ask for commitment"? Is safety a "priority" and a "condition of employment"? or Is safety a "value" and a "personal mission to actively care for the safety and health of others"?

Employees often consider themselves passive followers of safety rules and regulations. Why, because managers typically plan and evaluate most aspects of the job, including the safety protocol. As a result, the wageworker's perception of choice can be limited. Yet, an injury-free workplace requires interdependent engagement, information gathering, and AC4P coaching by the line workers. These are the employees who know most about the hazards and at-risk behaviors, as well as the factors contributing to these potential determinants of injuries and fatalities.

Perceived Competence

"People are not successful because they are motivated; they are motivated because they are successful" (Chance, 2008, p. 95). This provocative quotation reflects the powerful role of **perceived competence** in motivating people to continue working diligently and safely on a task when no one is watching or holding them accountable. Much research has shown that people become more self-motivated and self-directed when they feel competent at performing worthwhile work (Deci, 1975; Deci & Flaste, 1995; Deci & Ryan, 1995; Ryan & Deci, 2000). Offering genuine praise, recognition, gratitude, or **supportive feedback** can make that happen (Safety Objective 6). Unfortunately, our culture seems to downplay the value of praise for competent work, as illustrated in Figure 3.2.

Just the language we use when talking to ourselves (i.e., self-talk) and to others can influence perceptions of choice and/or competence. For example, which of the following pairs of words or phrases connect more to the perception of choice and/or competence and hence to self-motivation: "occupant restraint" or "vehicle safety belt," "requirement" or "opportunity," "peer pressure" or "peer support," "training" or "mentoring," "compliance" or "commitment," "I must meet this deadline" or "I choose to achieve another milestone."

FIGURE 3.2 Positive recognition builds competence.

Interdependence or a Sense of Community

Deci and Flaste (1995) affirm that we have three basic psychological needs, and when these needs are satisfied, we are self-motivated. Specifically, self-motivation is activated by conditions that facilitate the fulfillment of our needs for (a) autonomy or choice, (b) competence, and (c) relatedness—a sense of community or **interdependence.**

Consider these three popular slogans in American culture: "Nice guys finish last." "You've got to blow your own horn." "The squeaky wheel gets the grease." Now consider these quotations, popular in Japan: "One does not make the wind, but is blown by it" and "The nail that rises above the board invites a hammering down." These quotations reflect either an independent (individualistic) or an interdependent (collectivistic) mindset or perspective.

When coming into this world, people depend on others to take care of them. Children depend on parents or caregivers for all their basic needs. In contrast, adolescents look for opportunities to strike out on their own. As depicted in Figure 3.3, this reliance on self (**independence**) rather than others (interdependence) is promoted and supported throughout the U.S. culture, from high-school and college classrooms to corporate boardrooms.

High-performance teamwork requires a reciprocal dependency—team members depending on each other to complete their task assignments.

FIGURE 3.3 Teach children the power of interdependence.

This reflects a shift from independence to interdependence and a sense of community. With an interdependent community spirit, people trust each other to look out for their safety; and with self-motivation, individuals choose to contribute their competence for the safety and health of others in their work culture.

EMPOWERMENT

In the management literature, empowerment typically refers to delegating authority or responsibility. When managers say, "I empower you," they usually mean "Get'er done." As reflected in Figure 3.4, the message is, "Make it happen; no questions asked." In contrast, the AC4P safety leader first assesses whether the empowered individual *feels* empowered. A proper assessment of feeling empowered involves asking three questions derived from social learning theory (Bandura, 1982).

The first question is, "Can I do it?" Do I have the knowledge, skills, ability, and resources to achieve a particular objective? A "Yes" answer to this first empowerment question reflects a personal belief in having the competence to make it happen. Albert Bandura (1982) called this "self-efficacy."

Note that the term "self-efficacy" places the focus on personal belief. An observer might think an individual has the competence to complete a task, but the empowered individual might feel differently. Thus, a "Yes" answer to the first empowerment question implies a belief among those who received the assignment that they have the relevant personal competence to achieve the process or outcome goal.

FIGURE 3.4 Top-down control can stifle empowerment.

The second question—"Will it work?"—reflects **response efficacy**. Does the recipient of an assignment believe that performing the required behaviors will contribute to a valued mission for the performer and/or for others? In this case, education about the mission-based value of performing the task-relevant behavior may be needed. Regarding BBS, the response efficacy question translates into believing an AC4P behavioral coaching process (Safety Objective 6) will contribute to injury prevention and help to cultivate an AC4P work culture.

A negative answer to the self-efficacy question indicates a need for more training. A negative answer to the response efficacy question implies a need for education. People might believe they are able to accomplish a particular process or task (i.e., self-efficacy), but they might not believe such accomplishment will make a difference in a desired outcome (i.e., response efficacy). In this case, they need education, including an explanation of an evidence-based principle or theory, and perhaps the presentation of convincing data. Regarding the AC4P behavioral coaching process, substantial data are available to show that an AC4P behavioral observation-and-feedback process prevents workplace injuries (e.g., Geller, Perdue, & French, 2004; Sulzer-Azaroff & Austin, 2000).

The third question—"Is it worth it?"—targets motivation. Is the expected outcome worth the effort? Will performing the task-relevant behaviors result in a worthwhile outcome—achieving a positive consequence and/or avoiding a negative consequence? Figure 3.5 summarizes these three empowerment

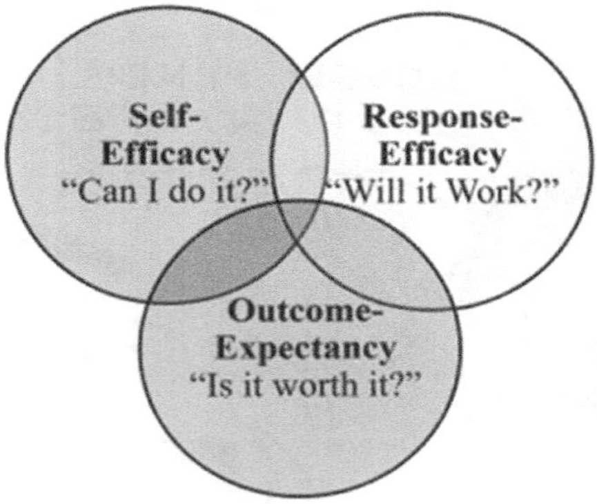

FIGURE 3.5 Three beliefs determine empowerment.

questions by referring to the three beliefs required to feel empowered: self-efficacy, response efficacy, and **outcome expectancy**.

After answering "Yes" to these three empowerment questions, meaningful behavior-focused goals need to be set. To make goal setting empowering, we propose the following SMARTS acronym: **S** for specific, **M** for motivational, **A** for attainable, **R** for relevant, **T** for trackable, and **S** for shared. We added the last "**S**" because sharing a behavioral goal with others enlists social support that can activate interpersonal reminders and behavioral feedback to facilitate goal attainment.

A **SMARTS goal** is empowering because it is attainable ("I can do it"), motivational ("It's worth it"), and relevant ("It will work"). Scott explains more about the connections between empowerment and self-motivation in a 15-minute TEDX talk on self-motivation (just google Scott Geller TEDX).

EMPOWERMENT VS. SELF-MOTIVATION

Although goal setting and empowerment precede the occurrence of behavior, each reflects the impact of motivational consequences. More specifically, feeling empowered means the individual has answered "Yes" to the motivational question, "Is it worth it?" and is activated to work toward achieving a given goal.

If the goal setting was SMARTS, the "M" for motivational implicates consequences. Feeling empowered implies the individual is ready or activated to work toward achieving a specified goal, which reflects the potential achievement of a particular consequence. Similarly, a self-motivated individual is anticipating or has received a consequence (e.g., recognition or supportive feedback) that supports self-directed rather than other-directed behavior.

Figure 3.6 illustrates how empowerment, vision, and goal setting fit with the activator-behavior-consequence (ABC) model of ABS. The simple but fundamental point: People need to feel empowered to work for goal achievement, including the anticipated acquisition of a desirable consequence and/or the expectation to avoid an undesirable consequence.

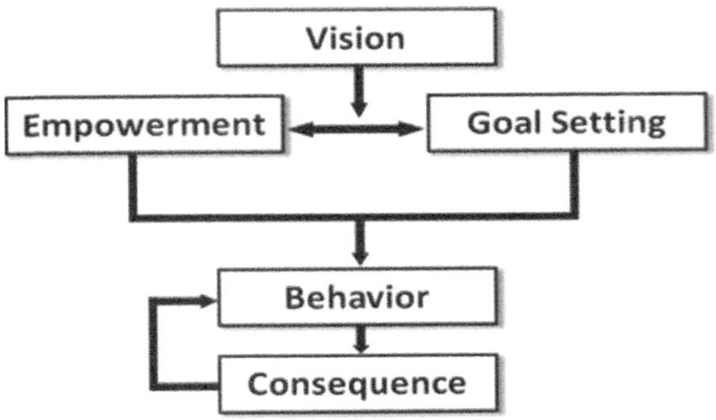

FIGURE 3.6 Empowerment and goal setting motivate with consequences.

People need to believe in and own the vision. They need to feel support from peers to attain process goals that support the vision by receiving supportive and corrective feedback to increase the quantity and improve the quality of behaviors consistent with vision-relevant goals. Note that behavioral consequences are crucial. Empowerment and goal setting can activate desirable behavior, but without relevant supportive consequences, the behavior will not last. It will extinguish, unless the consequences are intrinsic (Safety Directive 10).

Conclusion

The success of OHS processes increases as a function of the number of employees engaged in accomplishing them. People are more likely to initiate and sustain focused contributions to a process when they are self-motivated and feel empowered. Self-motivation increases when people perceive a sufficient degree of choice, competence, and community regarding a particular safety improvement intervention.

People feel empowered to contribute to an injury-prevention process when they perceive self-efficacy (“I can do it”) and response efficacy (“It will work"), and anticipate a beneficial outcome ("It's worth it"). In addition, behavioral goals are empowering when they are (a) challenging but attainable, (b) relevant to achieving a worthwhile mission, and (c) motivating because of anticipated consequences to gain or avoid.

While self-motivation, empowerment, and goal setting precede the performance of relevant behavior, each of these human dynamics includes an expectation of a desirable consequence and thereby activates motivation to perform.

DISCUSSION QUESTIONS

You can answer the following discussion questions from the information provided here for Safety Objective 3, but it would be useful at this point to watch a 15-minute TEDX talk on self-motivation by simply googling Scott Geller TEDX.

1. Discuss the difference between management and leadership with regard to real-world examples, and perhaps personal experience.

2. In what ways could management increase the perception of choice at your work facility with regard to OHS?

3. How might the perception of safety-related competence be enhanced among the workers at your facility?

4. How does your workplace culture support or stifle a sense of community or interdependence?

5. Explain the three beliefs that determine empowerment as a person-state, and suggest ways to increase these three empowerment beliefs in your work culture.

6. Distinguish between a goal and a vision with a personal example.

7. The ABC model of ABS (Safety Objective 2) presumes behavior is directed by activators and motivated by consequences. Given this perspective, how can goal setting provide both direction and motivation?

8. Select a safety directive of Safety Objective 3 and explain how that directive could be used to increase the number of employees engaged in supporting OHS at your work facility.

REFERENCES

Bandura, A. (1982). Self-efficacy mechanism in human agency. *American Psychologist, 37,* 122–147.

Bandura, A. (1997). *Self-efficacy: The exercise of control.* New York: W.H. Freeman and Company.

Chance, P. (2008). *The teacher's craft: The 10 essential skills of effective teaching.* Long Grove, IL: Waveland Press, Inc.

Deci, E. L. (1975). *Intrinsic motivation.* New York: Plenum.

Deci, E. L., & Flaste, R. (1995). *Why do we do what we do: Understanding self-motivation.* New York: Penguin Books.

Deci, E. L., & Ryan, R. M. (1995). *Intrinsic motivation and self-determinism in human behavior.* New York: Plenum.

Geller, E. S. (2016). The psychology of self-motivation. In E. S. Geller (Ed.), *Applied psychology: Actively caring for people* (pp. 83–118). New York: Cambridge University Press.

Geller, E. S., Perdue, S., & French, A. (2004). Behavior-based safety coaching: Ten guidelines for successful application. *Professional Safety, 49*(7), 42–49.

Reed, D., et al. (2016). Actively caring for higher education. In E. S. Geller (Ed.), *Applied psychology: Actively caring for people* (pp. 563–593). New York: Cambridge University Press.

Ryan, R. M., & Deci, E. L. (2000). Self-determination theory and the foundation of intrinsic motivation, social development, and well-being. *American Psychologist, 55,* 68–75.

Sulzer-Azaroff, B., & Austin, J. (2000). Does BBS work? Behavior-based safety and injury reduction: A survey of the evidence. *Professional Safety, 45*(7), 19–24.

4 Safety Objective 4
Conduct Optimal Analyses of Close Calls, Injuries, and Fatalities

Safety Directive 19: Consider the nature/nurture interaction.
Safety Directive 20: Consider how consequences motivate behavior.
Safety Directive 21: Consider the power of observational learning.
Safety Directive 22: Consider the Fundamental Attribution Error.
Safety Directive 23: Consider the Self-Serving Bias.
Safety Directive 24: Beware of Groupthink.

This safety objective addresses the challenge of conducting an unbiased, acceptable, and useful analysis of a close call, property damage, or personal injury. We do not use the term "investigation," although that term is commonly used in the safety world. Since language affects attitude, behavior, and culture, we object to using a term with negative connotations that can decrease interpersonal involvement in a most critical component of injury prevention.

Employees can interpret an "investigation" as an interrogation to find someone to blame, as in a "criminal investigation." Injury investigations are often conducted to find a "root cause," and this can be perceived as attributing the "cause" to an individual.

Employees should not view a safety-related "analysis" as an attempt to find the cause of an injury. An analysis is conducted to define environmental, behavioral, and personal factors—leading indicators—that may have contributed to a close call or an injury. If contributing factors are identified and changed appropriately, the probability of a more severe injury occurring in the future will be decreased. Thus, an accurate and accepted "incident analysis" is critical for injury prevention.

THE NATURE/NURTURE INTERACTION

As depicted humorously in Figure 4.1, people's behavior results from an ongoing interaction between nature and nurture. Contemporary psychology—the

DOI: 10.1201/9781003758822-5

science of human experience—addresses the impact of both personality **traits** (i.e., innate dispositions) and situational factors (e.g., behavioral consequences) that influence human behavior, attitudes, and emotions. The significant impact of nature or heredity on safety-related behavior cannot be denied, but safety leaders need to focus on factors they can control to manage risk and prevent injuries.

Two personality traits—conscientiousness and sensation seeking—significantly impact people's behavior. A conscientious individual is organized and detail-focused and inclined to be rule-governed (e.g., a follower of specified safety regulations), and individuals scoring high on the Zuckerman Sensation Seeking Scale (Zuckerman, 1979) are more likely to take risks. Does this mean a less conscientious sensation seeker is more likely to be injured? Probably not, because of the dramatic impact of situational factors. For this reason, numerous studies designed to identify the "injury-prone" personality have been unsuccessful (Geller, 2005).

FIGURE 4.1 The nature/nurture interaction affects behavior.

It is interesting to consider natural tendencies to take risks and/or avoid injury-prevention behavior, but it is more practical to treat a personality **disposition** as a "state" rather than a "trait," and to entertain the impact of nurture over nature. Regarding conscientiousness, for example, situational factors can increase an individual's self-motivation, persistence, and rule following, in spite of a relatively low score on an assessment of conscientiousness.

Plus, situational factors (e.g., a safety-related intervention) can influence people with a proclivity to seek excitement and take risks to use personal protective equipment (PPE). Thus, the systematic analysis of a close call or an injury must focus on nurture—the situational factors that can be improved to reduce the probability of a hurtful interaction between an individual's behavior and the work environment.

THE FUNDAMENTAL ATTRIBUTION ERROR

Figure 4.2 illustrates the **Fundamental Attribution Error** that probably biases every evaluation or analysis of a worker's close call or injury. Specifically, much research and most people's commonsense support the tendency for individuals to automatically attribute the observed behavior of another person to that individual's nature, character, innate abilities, personality traits, or in one word—disposition (Heider, 1958).

FIGURE 4.2 The caddie is making the Fundamental Attribution Error.

We overemphasize other people's internal attributes to explain their behavior, and we often ignore the fact that situations play a significant role in determining what people do. This **attributional bias** is called "fundamental" because it is virtually universal, and it is an "error" because such judgments are most often wrong. Simply put, the Fundamental Attribution Error reflects an overemphasis on nature over nurture.

Many readers have likely experienced this bias when considering or talking about the cause of an injury to an employee. Supervisors or safety professionals are apt to blame characteristics of the worker for a mishap. Failing to consider the influence of the situation or context, they think or say aloud, "He was careless or risky."

Indeed, a primary criticism of BBS has been the inaccurate presumption that BBS blames workers for their injuries without considering the broader context of situational factors (Geller, 1997; Hans, 1996; Manuele, 1998). In fact, we have witnessed a consultant promote the BBS approach at several national safety and health conferences by showing video clips of workers performing exceedingly risky behavior and then claiming the need for BBS while the audience laughs at the silly antics shown on the screen.

Note that the golfer in Figure 4.2 is not blaming himself, but attributing his poor performance to the environment—wind, high grass, and inadequate equipment. This focus on situational determinants of one's own behavior reflects another error of attribution—the **Self-Serving Bias**. Let's consider how this human dynamic can make an injury analysis less helpful or constructive.

THE SELF-SERVING BIAS

As illustrated in Figure 4.3, the Self-Serving Bias is an attributional process in which people explain their own behavior—successful or unsuccessful—for personal advantage (Mezulis et al., 2004). In other words, people explain their performance to enhance or protect their self-esteem. We maintain positive views of ourselves by blaming bad outcomes on features of the situation.

On the other hand, when things go well, we attribute the success to our own skills, abilities, and/or positive qualities of our personality. Readers who have been questioned about a personal error that resulted in property damage and/or an injury have probably experienced this attributional bias.

FIGURE 4.3 An injured customer displays the Self-Serving Bias.

A Self-Serving Bias in moderate doses can be useful because it protects self-esteem; and with an analysis of a close call or an injury, a focus on situational factors can lead to environmental adjustments capable of keeping workers safe. However, if people consistently deny or resist the role of dispositional attributes or personal responsibility for an observation of subpar (e.g., at-risk) behavior, they stifle considerations and opportunities for personal improvement.

This raises the critical challenge of developing the kind of interdependent work culture in which employees have the *courage* to offer coworkers corrective behavioral feedback, the *humility* to accept corrective feedback from others, and the *integrity* to make suitable behavioral adjustments. This happens when people give supportive feedback well and more often than corrective feedback, and when they deliver corrective feedback effectively, as addressed in Safety Objective 6.

The last directive of Safety Objective 4 illustrates the undesirable outcome of a culture that hinders the delivery of interpersonal behavioral feedback for personal and group improvement.

BEWARE OF GROUPTHINK

Groupthink refers to a tendency for groups to reach a quick decision without taking the time for substantial feedback, discussion, and consensus-building (Janis, 1972, 1983). Sometimes the supervisor or manager of a group (e.g., a work team) discourages disagreement and advocates unanimity in order to make an efficient decision and move a process along. In that context, individuals may be unlikely to challenge peers or colleagues for fear of losing status or a friendship. The result: a deterioration of thoughtful decision-making, improvement considerations, and moral judgment.

Groupthink decreases the possibility of **synergy**—a process of interdependent participation in a goal-directed activity that results in greater achievement than is possible from all participants performing alone and independently (Safety Objective 7).

Figure 4.4 illustrates Groupthink as typically explained. The group members disagree with the decision, but they do not have the courage to speak up and question or challenge the manager's perspective. Group leaders decrease the probability of Groupthink when they embrace diverse opinions, invite input and critique, and challenge individuals to "think outside the box." When leaders admit vulnerability, own up to their mistakes, and solicit corrective feedback, they set the stage for continuous improvement and naturally stifle Groupthink.

FIGURE 4.4 Groupthink stifles individual contribution.

While Groupthink is not typically associated with analyzing the contributing factors to a close call, injury, or property damage, the connection should be obvious. The usefulness of an analysis and a report of an occupational mishap usually depend on the number of individuals who provide frank and open information from their relevant personal experience. Beyond the person injured, this should certainly include others who have worked on the same or a similar job.

Imagine a work culture in which everyone familiar with the situation in which a close call or an injury occurred speaks up to offer feedback relevant to preventing future occurrences of a similar incident. Widespread and effective engagement of people giving their best efforts for the safety of themselves and others requires leadership that inspires safety-related empowerment and self-motivation throughout a work culture (Safety Objective 3), as well as effective behavior-based actively caring for people (AC4P) coaching (Safety Objective 6).

Conclusion

Safety Objective 4 explained and illustrated three evidence-based phenomena from psychological science that can decrease the acceptance and benefits of analyses of close calls, injuries, and property damage—the Fundamental Attribution Error, the Self-Serving Bias, and Groupthink.

In addition, the interaction of dispositional and situational determinants of behavior was discussed in order to illustrate how nurture or personality *states* can overpower nature or personality *traits*. For this to happen, however, an interdependent AC4P work culture needs to be cultivated—one in which Groupthink is nonexistent and all employees feel empowered and self-motivated to speak up on behalf of the safety of themselves and others. Such a work culture is considered "psychologically safe" (Clark, 2020; Geller, 2022).

Effective organizational leaders promote **psychological safety** by helping employees: (a) feel they are included; (b) believe they are contributing to the organization's mission, (c) are learning from their own and other's contributions, and (d) can challenge the status quo (Clark, 2020).

DISCUSSION QUESTIONS

1. Discuss the role of human *nature* versus *nurture* with regard to an injury or an at-risk behavior you have experienced or observed.

2. How do consequences influence your performance of safe versus at-risk behavior?

3. What work-related behaviors have you learned by observing the behaviors of other workers? Which of those behaviors were safe or at risk?

4. Distinguish between the Fundamental Attribution Error and the Self-Serving Bias with one or more real-world examples.

5. Have you ever experienced Groupthink? Please explain.

6. Suggest some practical ways for stifling the occurrence of Groupthink and promoting psychological safety at your work facility.

7. Which of the safety directives for Safety Objective 4 could influence a more useful and unbiased analysis of an injury at your facility? Please explain.

REFERENCES

Clark, T. R. (2020). *The 4 stages of psychological safety: Defining the path to inclusion and innovation*. San Francisco, CA: Berrett-Koehler Publishers.

Geller, E. S. (1997). What is behavior-based safety, anyway? *Occupational Health & Safety*, *66*(1), 25–28.

Geller, E. S. (2005). *People-based safety: The source*. Virginia Beach, VA: Coastal Training and Technology Corporation.

Geller, E. S. (2022). Psychological safety: The optimal context for injury prevention. *Professional Safety, 67*(1), 18–20.

Hans, M. (1996). Does behavior-based safety work? *Safety and Health*, National Safety Council, June, 44–49.

Heider, F. (1958). *The psychology of interpersonal relations*. New York: Wiley.

Janis, I. L. (1972). *Victims of groupthink*. Boston, MA: Houghton-Mifflin.

Janis, I. L. (1983). *Groupthink: Psychological studies of policy decisions and fiascoes* (2nd ed.). Boston, MA: Houghton-Mifflin.

Manuele, F. A. (1998). Perspectives on behavioral safety: Observations of ASSE's behavioral safety symposium. *Professional Safety, 43,* 32–37.

Mezulis, A. H., Abramson, L. Y., Hyde, J. S., & Hankin, B. L. (2004). Is there a universal positivity bias in attributions? A meta-analytic review of individual, developmental, and cultural differences in the self-serving attributional bias. *Psychological Bulletin, 130*(5), 711–747.

Zuckerman, M. (1979). *Sensation seeking: Beyond the optimal level of arousal.* Hillsdale, NJ: Lawrence Erlbaum.

5 Safety Objective 5
Apply Social Influence Principles to Enhance OHS Engagement

Safety Directive 25: Apply the Principle of Consistency.
Safety Directive 26: Apply social labeling and the hypocrisy effect.
Safety Directive 27: Apply the Principle of Reciprocity.
Safety Directive 28: Use the door-in-the-face technique.
Safety Directive 29: Apply the Principle of Authority.
Safety Directive 30: Apply the Principle of Scarcity.

We are social animals. Indeed, our interpersonal social relationships define us—our perceived competence, our sense of belonging, and our self-esteem. We participate with others on a daily basis to achieve personal and group goals, and those interactions can enhance or stifle our motivation to keep on keeping on. Safety Objective 5 addresses critical research-based principles of social influence that can facilitate or hinder employee engagement for the prevention of workplace injuries.

The social dynamics of an organization both reflect and affect its culture (Safety Objective 10). Certain qualities of a work setting influence interpersonal social dynamics, and these in turn influence how a work culture dramatically influences OHS. The social influence principles explained and illustrated here enable an appreciation of the personal and interpersonal dimensions of a culture, and suggest strategies for improving the social dynamics of a work culture. We begin with an explication of the powerful **Principle of Consistency**, which research suggests is generally the most popular and clearly relevant for enhancing employee engagement for OHS (Geller, 2002).

THE PRINCIPLE OF CONSISTENCY

The need for congruence between what we think, how we feel, and what we do fuels this social influence principle. We experience tension or **cognitive**

DOI: 10.1201/9781003758822-6

dissonance (Festinger, 1957) when perceiving a lack of consistency between a behavior we *choose* to perform and another behavior, attitude, or personal value. The key word in that sentence is *choose*. When we feel coerced into doing something, we do not feel uncomfortable if that behavior is inconsistent with a personal attitude, value, or another behavior.

The Foot-in-the-Door Technique

As illustrated humorously in Figure 5.1, an individual who chooses to follow a small request is likely to comply with a larger request later. For example, after agreeing to serve on a safety steering committee, an individual is likely willing to give a presentation at a safety meeting. Researchers call this the **foot-in-the-door technique**, and they have found it to be quite successful at boosting product sales, monetary contributions to charities, and blood donations (Cialdini, 2001; Freedman & Fraser, 1966).

However, this technique only works if the individual complies with the initial small request. If the person says "No" to the first request, this individual will actually find it easier to refuse a second, more important request. In this case, the Principle of Consistency is working against the requester.

FIGURE 5.1 A foot in the door leads to more.

Social Labeling

The **Principle of Consistency** is responsible for the power of a **social label** which is evident when an individual is assigned an ability, attribute, attitude, or belief label. In other words, people attempt to live up to the labels they are given when situations and social dynamics support that label. People want to appear consistent. For example, labeling a driver as careful and safe can increase the frequency of his/her safe driving behaviors. In contrast, the label "reckless driver" can prompt the driver to take more risks behind the wheel.

The seminal research by Carol Dweck (2006) indicates the need to use an *effort* or behavior-based label (e.g., "She tries hard.") rather than an *ability* label (e.g., "She is gifted.") for a desirable characteristic. Why, because labels that imply a fixed ability rather than growth potential can stifle one's motivation to take on challenges. For example, an individual with a fixed-ability label is likely to resist volunteering for a challenging project with such self-talk as, "If I accept this challenge and fail, people's positive perceptions of my special ability will decrease."

So how should you recognize an employee for giving a passionate safety message at a toolbox meeting? Ability-oriented acknowledgment would focus on the employee's natural talent to deliver an engaging presentation—"Great safety talk Sam, you're a born speaker with innate leadership qualities." Such nonbehavioral feedback would certainly make Sam feel good and would surely increase his self-esteem. But that fixed-ability feedback does not support a particular behavior or motivate continuous improvement.

This type of feedback could stifle Sam's willingness to attempt a more challenging speaking opportunity with self-talk like

> They expect me to give that presentation with limited preparation because they presume I have an innate ability to speak in front of people. But in the past, friends and coworkers were my audience, not the strangers they want me to address this time. I am so nervous about that assignment because I might not live up to their high expectations.

In contrast, effort-based recognition would pinpoint both the commendable behavior observed and the motivation displayed—"Thanks for pointing out the safe behaviors you observed on the job, Sam. Your passion for safety is obvious, and your commitment to learning more and more about the AC4P Movement is much appreciated, thank you so much." Note that this effort-based supportive feedback specifies the behavior deserving the appreciation and reflects motivation to learn and continuously improve. In addition, it ends with an expression of sincere gratitude.

The Hypocrisy Effect

This behavior change intervention uses the Principle of Consistency to motivate the occurrence of a desirable target behavior. The change agent (e.g., a safety professional or supervisor) asks participants to think about their *ideal* occurrences versus their *actual* performance of a particular desirable behavior in order to activate a perceived behavioral inconsistency. This inspires more occurrences of the desired behavior (Aronson, 1999). Consider the following three steps to increase the frequency of a safety-related behavior:

1. Give one or more participants a convincing rationale for a particular safety-related behavior (e.g., "When you're driving a vehicle with passengers, you should always remind everyone to buckle up.").
2. Ask the participant(s) to commit publicly and actively to perform this safety-related behavior whenever it is warranted by the situation.
3. Ask the participant(s) to list the most recent times they did not perform the particular safety-related behavior of their chosen commitment.

In a sense, this technique activates a "guilt trip." Most participants will recognize an inconsistency between their prior behavior and a current belief or conviction that led to their current behavioral commitment. As mentioned earlier, psychologists call such awareness of a behavior-attitude discrepancy *cognitive dissonance* (Festinger, 1957). Whatever the label, this process reflects an effective application of the Principle of Consistency—"thinking a person into a certain action or behavior."

THE PRINCIPLE OF RECIPROCITY

The phrase "Do for me and I'll do for you" reflects the Principle of Reciprocity of social influence (Cialdini, 2001). If you are nice to someone, they will feel obligated to do something nice for you or perhaps for another person. Many charitable organizations include a small gift in their solicitation packages (e.g., address labels, a calendar, a pen, or a small writing pad), assuming the recipient will feel obligated to return the favor with a financial donation.

For OHS, this means safety leaders should look for opportunities to go out of their way for another person's safety. When individuals actively care for someone else's safety, they set the stage for the **Principle of Reciprocity**—they increase the likelihood that the recipients of safety-related caring behavior will actively care for the safety of someone else.

Systems Thinking

Systems thinkers understand that many interdependent situational and dispositional factors influence behavior, and the occurrence of a particular behavior can affect the environmental factors and person states of a system. They realize that their reaction to people who do them a favor can either stifle or mobilize a spiral of interpersonal reciprocity.

For example, when systems thinkers receive a "Thank You" for their safety-related behavior, they do not demean the favor by saying something like "No problem" or "It was really nothing." This can make the safety-related behavior appear trivial and reduce the impetus for follow-up reciprocity or pay-it-forward behavior.

Instead, systems thinkers react to a "Thank You" with something like "You are very welcome, but I know you'd do the same for me." This reply supports the Principle of Reciprocity and perhaps increases the occurrence of more pay-it-forward safety-related behavior.

The Door-in-the-Face Technique

Suppose the plant safety director asked you to chair the safety steering committee for the next two years. Let's assume you perceive this request as outrageous, given your other assignments and the fact you have never even served on such a committee. You reply, "Thanks for asking, but no thanks." The safety director says he understands, and then asks whether you would be willing to serve as an active member of the safety committee. According to psychological science, because the safety director retreated from his initial request, you will feel subtle pressure to make a similar concession—you reciprocate—and agree to the less demanding request (Cialdini, 2001; Cialdini et al., 1975).

Why do lawyers ask for outlandish amounts of money at the start of a civil trial? Why do labor negotiators start with extreme demands? Perhaps these agents of influence do not expect to receive their initial request, but they have learned that they are more likely to succeed with a second more realistic request after retreating from the first. Through real-world negotiation, these professionals have learned the power of the reciprocity-based **door-in-the-face technique**. This social influence strategy is illustrated humorously in Figure 5.2.

FIGURE 5.2 Ask for a lot but settle for less.

THE PRINCIPLE OF AUTHORITY

From childhood through adulthood, we learn to follow legitimate authority—from "mother knows best" to "the boss knows best" (Milgram, 1974). That gives us an excuse to avoid taking personal responsibility for our actions. If someone with authority asks employees to take a risk, they might be very willing to comply because if something goes wrong, it will not be their fault. They can blame the person who told them to take the risk. It is easy to see how this relates to workplace safety. Safety leaders need to be aware of the Principle of Authority and encourage people to resist the temptation to follow orders blindly and mindlessly.

Imagine that a supervisor requests you to avoid a particular time-consuming OHS process—such as a lockout-tagout protocol, a site-specific safety plan, or a confined space inspection—in order to get a certain job done quickly. Would you immediately comply with this request or would you question these instructions from an authority figure at your workplace? Note how your answer to that rhetorical question is likely influenced by psychological safety, as introduced in the prior chapter—Safety Objective 4.

The term "authority" does have negative connotations because many historical examples and research in psychological science have illustrated the top-down coercive influence of people abusing their power of authority, as exhibited during war and tragic genocides such as the Holocaust.

In the 1960s, Stanley Milgram and associates demonstrated that 65% of intelligent and well-meaning college students followed orders to administer a 450-volt electric shock to a screaming peer (Milgram, 1963). Even Milgram was surprised that so many of his subjects, ranging in age from 20 to 50, obeyed the experimenter's "shocking" requests, concluding that "Ordinary people, simply doing their job, and without any particular hostility on their part, can become agents in a terrible destructive process" (Milgram, 1974, p. 6). Relatedly, workers might overlook obvious safety hazards and put themselves and others at risk with the self-talk, "I'm just following orders."

On the other hand, many people with authority (e.g., physicians, ministers, teachers, parents, and safety professionals) activate desirable behavior. People use safety devices, take vitamin pills and supplements, follow security precautions, and get periodic medical exams because one or more persons with credible authority told them to perform those behaviors. Even though we rarely perceive an immediate health or safety benefit from complying with such injury prevention or health promotion advice, we take the word of an authority figure that such behavior will eventually pay off. Hence, the Principle of Authority can have beneficial or debilitating consequences to human welfare, depending on who has the perceived authority.

FIGURE 5.3 Mandates can be counterproductive.

THE PRINCIPLE OF SCARCITY

Have you ever gone out of your way to purchase front-row tickets to a sporting event or concert? Have you jammed yourself into a department store to get a "limited-time" bargain? Or have you participated at an auction when a one-of-a-kind item was displayed and sold to the highest bidder? These situations and the rhetorical questions illustrate how the value of something can increase with perceived scarcity.

When Individuality Is Threatened

As illustrated in Figure 5.3, when top-down control makes individuality or the perception of **personal control** scarce, some people will exhibit contrary behavior. Why? Behavior contrary to a mandate enables a regaining of personal choice or freedom. Social psychologists refer to this phenomenon as "psychological reactance" (Brehm, 1966), while behavioral scientists call it "countercontrol" (Skinner, 1971). Regardless of the label, reacting to show individuality can be extremely counterproductive to OHS.

Scott once met a worker who wore safety frames—not safety glasses, only safety frames—because he had removed the lenses. When his supervisor walked down the aisle, this employee would look right at him and wave. This employee's coworkers actually lauded him for thumbing his nose at the safety system. At this facility, the employees perceived safety as a top-down mandate that restricted their individual freedom. This particular worker actually increased his status in his work culture by demonstrating psychological reactance or countercontrol.

Bottom line: A command-and-control approach to OHS can make personal freedom scarce and hinder participation for safety. When people perceive their work culture as restrictive, they may attempt to beat the system. The **Principle of Scarcity** indicates that true commitment and long-term engagement for OHS cannot be dictated (Geller, 2024).

Fear of Losing

It can be very beneficial to tell audiences at safety workshops that their best opportunity to implement the principles and procedures they learned is now—while the material is fresh and their organization expects them to share and apply what they have learned. As time passes, opportunities and confidence to apply new information fade, because one's perceived competence or self-efficacy can diminish over time (Bandura, 1997). And as time passes, management's expectancy and obligation to consider a change can get blurred by the familiar routine.

This application of the Principle of Scarcity focuses on what an individual could lose by failing to take advantage of a one-of-a kind opportunity. Social psychologists have shown that people are especially motivated to avoid a loss (Cialdini, 2001). When the need to act immediately to avoid a loss is evident, people are aroused to mindful action. The power of the Principle of Scarcity to motivate participation in safety-related activities is evident when the potential of loss is activated (as illustrated in a personal injury testimony), and when loss-avoidance strategies are available.

For ten consecutive keynote presentations at the annual National Safety Council Congress and Expo, Scott teamed up with Charlie Morecraft to take advantage of the Principle of Scarcity. Charlie spoke first for an hour and explained with remarkable passion and emotion the severe and long-term losses he experienced following a tragic incident at an Exxon site on August 8th, 1980 (8/8/80). He ended each of his dynamic presentations with the assertion that now the audience of safety leaders knows *why* they do what they do to prevent workplace injuries. Then he introduced Scott by asserting, "Now, Dr. Geller will explain *how* you can prevent the kind of losses that I have endured."

After their first five keynote addresses, Krista stepped up with notable passion and commitment for OHS to introduce Charlie and Scott for their remaining keynotes. That experience launched Krista's career as a knowledgeable and inspirational keynote speaker and workshop instructor at OHS conferences nationwide and beyond.

Conclusion

Achieving an injury-free workplace requires nothing less than a culture transformation. Every participant at the workplace must grow from a dependent and independent mindset to interdependency. Employees cannot rely on their company to keep them safe solely through engineering intervention. Nor is it sufficient for employees to count on their own individual efforts to keep themselves injury-free. Rather, workers need others to remove environmental hazards they miss and provide corrective feedback for their at-risk behavior (Safety Objective 6). This interdependency requires ongoing interpersonal communication.

The social dynamics of a culture determine whether constructive interpersonal exchanges are likely to occur and whether the impact of these interactions will benefit or damage safety. The social influence principles described here reveal basic social dynamics that can facilitate or inhibit OHS.

People's desire to be consistent influences their resistance to change. However, when they choose to change even slightly, the Principle of Consistency can facilitate more commitment and more beneficial change. With this principle in effect, even small increases in participation can result in supportive attitude change, followed by more participation and more desirable attitude change.

When workers actively care for the safety of another person, they induce a sense of obligation to return the favor to that person and/or to others—the Principle of Reciprocity. They also set an example that spreads more AC4P behavior throughout the workplace—the *observational learning* of Safety Objective 4.

The Principle of Authority indicates that people may follow orders blindly and mindlessly. This becomes a barrier to OHS when a manager or supervisor asks a worker to perform a task that involves at-risk behavior. However, those in positions of authority can have the opposite effect. Their positive role-modeling behavior and visible ongoing support of interdependency and actively caring are key to achieving an injury-free workplace.

Finally, the Principle of Scarcity explains why some employees actively resist complying with top-down safety rules, regulations, or a quick-fix safety program. In some command-and-control work cultures, noncompliance or nonparticipation reflects a personal statement of "freedom." Employee engagement is much more likely with top-down support of safety processes that are developed, owned, and continuously improved by employee work teams that understand the rationale and relevancy of the safety process.

Highlighting the uncommon and rare aspects of a safety process uses the Principle of Scarcity to fuel participation. Involvement in a new safety effort can be activated by explaining that the "window of opportunity" for implementation is brief, and specifying what can be lost if substantial participation is not forthcoming as soon as possible.

DISCUSSION QUESTIONS

1. Explain how you could apply the Hypocrisy Effect to increase occurrences of a particular safety-related behavior.

2. How does the Principle of Consistency influence the Hypocrisy Effect?

3. Do you think the Principle of Reciprocity is as influential today as it was in the 1970s when most of the research of this social influence principle was accomplished? Why or why not?

4. A common reply following a "Thank You" expression of gratitude is: "No problem." Why is "You're very welcome" a better reply?

5. Provide some real-world examples of the "foot-in-the-door" and the "door-in-the-face" techniques.

6. How can the Principle of Authority be both good and bad for OHS?

7. How does a perceived loss of personal freedom or a sense of individuality stifle employee engagement in OHS sometimes?

8. Provide some examples of independence and interdependence with regard to OHS at your workplace?

REFERENCES

Aronson, E. (1999). The power of self-persuasion. *American Psychologist, 54*(11), 875–884.

Bandura, A. (1997). *Self-efficacy: The exercise of control.* New York: W. H. Freeman.

Brehm, J. W. (1966). *A theory of psychological reactance.* New York: Academic Press.

Cialdini, R. B. (2001). *Influence: Science and practice* (4th ed.). Needham Heights, MA: Allyn & Bacon.

Cialdini, R. B., Vincent, J. E., Lewis, S. K., Catalan, J., Wheeler, D., & Darby, B. L. (1975). Reciprocal concessions procedure for inducing compliance: The door-in-the-face technique. *Journal of Personality and Social Psychology, 31*, 206–215.

Dweck, C. S. (2006). *Mindset: The new psychology of success.* New York: Ballantine Books.

Festinger, L. (1957). *A theory of cognitive dissonance.* Stanford: Stanford University Press.

Freedman, J. L., & Fraser S. C. (1966). Compliance without pressure: The foot-in-the-door technique. *Journal of Personality and Social Psychology, 4*, 195–203.

Geller, E. S. (2001). *The psychology of safety handbook.* Boca Raton, FL: CRC Press.

Geller, E. S. (2002). *The participation factor: How to increase involvement in occupational safety.* Des Plaines, IL: American Society of Safety Engineers.

Milgram, S. (1963). Behavioral study of obedience. *Journal of Abnormal and Social Psychology, 67*, 371–378.

Milgram, S. (1974). *Obedience to authority: An experimental view.* New York: Harper & Row.

Skinner, B. F. (1971). *Beyond freedom and dignity.* New York: Knopf.

6 Safety Objective 6
Apply Interpersonal Coaching to Improve OHS Behavior

Safety Directive 31: Teach and apply the COACH acronym—Care, Observe, Analyze, Communicate, and Help.
Safety Directive 32: Make feedback behavioral.
Safety Directive 33: Use more supportive than corrective feedback.
Safety Directive 34: Promote success-seeking over failure-avoiding.
Safety Directive 35: Communicate with empathy.
Safety Directive 36: Live by the Platinum Rule, not the Golden Rule.

Interpersonal behavior-based coaching is essential for any mission to keep people safe. In fact, the success of BBS or people-based safety (PBS) is contingent on the implementation of an effective peer-to-peer coaching process. One peer (the observer) uses an employee-derived critical behavioral checklist (CBC) to observe and record the work process of another peer. Potential environmental determinants of at-risk behavior and inhibitors of safe behavior are noted in a "comments" column of the CBC.

When a CBC is completed, a percent-safe score is calculated and entered into computer software for a comparative analysis of safe behavior percentages across work teams, job sites, and company facilities. Software helps organize and summarize the results from companywide CBCs and pinpoints targets for corrective action. This data analysis component of BBS is critical for proactive injury prevention—but this is not the most important safety success feature of BBS.

Most records of behavioral observations are biased and unreliable (Geller, Perdue, & French, 2004). Why, because CBC records are typically obtained under unnatural conditions, as when the observations are announced beforehand. There is also a tendency for peers to overlook at-risk behavior of coworkers, especially when an interpersonal feedback session is expected to follow the behavioral observations.

DOI: 10.1201/9781003758822-7

CBC data provide useful leading indicators, but you should not take the absolute value of those numbers too seriously. Actually, the most powerful feature of a peer-to-peer observation and feedback process is the very component many organizations implement ineffectively or actually omit entirely—interpersonal behavior-based coaching.

BEHAVIOR-BASED COACHING

The letters of COACH say it all. Interpersonal coaching for safety begins with **C**aring. This is not a "gotcha" process focused on finding faults or mistakes from others. It is an AC4P process. People acknowledge and support the safe behavior of a coworker and strategically point out opportunities for improvement. This is the critical corrective feedback component of coaching for safety, and we explain this communication process below.

"When you know I care, you will care what I know. In fact, I care so much I'm willing to observe your behavior—with your permission of course—and offer behavior-focused feedback." This quotation reflects the **O**bserve phase of AC4P coaching. As mentioned above, observers use a CBC to record observations of safe and at-risk behavior, as well as the possible environmental determinants of those behaviors.

Observing and recording the ongoing interaction of specific behaviors and the environmental/cultural conditions that facilitate at-risk behavior or inhibit safe behavior is the **A**nalyze component of the COACH acronym.

Next, we have another "**C**" for "**C**ommunicate"—showing sincere appreciation and gratitude for the safe behaviors observed and delivering corrective feedback for occurrences of at-risk behavior.

When the person observed perceives this communication phase to be constructive, especially by accepting and owning the corrective feedback, the last letter of COACH reflects the outcome—**H**elp. Let's consider strategies to make the communication phase of AC4P coaching most helpful.

Make Feedback Behavioral

Obviously coaching feedback should be behavioral, but feedback is too often nonbehavioral. While "Nice job," "Thank you for supporting our team," and "Excellent presentation" are supportive and pleasing to hear, these statements are not associated with desirable behavior and are not as beneficial as they could be. When people know what they did to earn sincere appreciation or praise, they are more likely to perform that behavior again.

What about corrective feedback? Your personal experience or commonsense likely reflects the systematic field observations of athlete coaches. Specifically, Carter et al. (2017) observed coaches on several NCAA Division I sports teams and noted that all coaches gave more corrective than supportive feedback. While supportive feedback was more often nonbehavioral (e.g., "Outstanding effort!"), corrective feedback often specified the undesirable behavior observed and frequently included a corrective action strategy (e.g., "Try swinging the bat in a more level and fluid motion.").

Athletes will accept and practice behavioral adjustments offered by a respected coach because they want to improve a skill they perceive to be important. However, receiving corrective feedback about an at-risk behavior observed in a work setting is a different story.

Figure 6.1 illustrates humorously a parent's attempt to be positive about an undesirable outcome. In this case, corrective behavioral feedback is called for. As indicated above, it is important to be directive when giving supportive feedback, but crucial to be **nondirective** when offering correction. While supportive feedback includes a specification of the desirable behavior you observed, when giving corrective feedback it is best to ask questions first.

FIGURE 6.1 Dad's attempt to be positive misses the mark.

With corrective feedback for OHS, the objective is to get the feedback recipient to accept the observation of an at-risk behavior, and then state a sincere intention to improve. This is more likely to happen if the observer shows genuine intent to learn the perspective of the person s/he observes working at risk. It is essential to listen attentively to explanations or excuses for not following a safety protocol. This could uncover situational factors that motivated or facilitated performance of the at-risk behavior and/or inhibited an occurrence of the safe alternative.

In the case of the undesirable report card in Figure 6.1, Dad should ask his daughter to identify behaviors that led to the low grades, and then to specify behaviors that could improve the next report card. If Dad listens actively to his daughter's explanation for the low grades, as well as her own suggestions for improvement, he might receive the ultimate reward from behavior-based coaching. His daughter asks him for advice and/or support in achieving her behavioral goals and improving her grades. Of course, ownership of undesirable performance and a commitment to improve are more likely if employees follow the next commonsense coaching strategy throughout their work culture.

Use More Supportive than Corrective Feedback

Question: If your boss or work supervisor asks you to come to his/her office for some "feedback," how do you feel? Do you feel good, anticipating some positive recognition for certain notable behavior, or do you expect a reprimand for an undesirable behavior? Answers to these questions likely attest to an unfortunate negative connotation of the term "feedback." Negative or corrective feedback is typically given more often than positive or supportive feedback. Figure 6.2 illustrates this critical large-scale problem with the common perception of "feedback"—a necessary component of coaching for safety success.

FIGURE 6.2 Feedback often implies a negative consequence.

"We learn more from our mistakes." How many times have you heard that? The statement might allow us to feel better about the errors of our ways and provide an excuse for focusing more on people's failures than on their successes, but nothing could be further from the truth. Behavioral scientists have shown convincingly that success—not failure—produces the most effective learning (Chance, 2008; Reed et al., 2016; Thorndike, 1931).

Supportive behavioral feedback not only maintains or increases occurrences of the desirable (e.g., safe) behavior that is recognized, but it promotes a positive mindset or disposition that can fuel optimism and self-motivation (Safety Objective 3). Optimists respond to setbacks in a positive and adaptive manner, and are willing to accept challenges, as opposed to evading demanding tasks in order to avoid failure (Seligman, 1991). Geller and Geller (2017) refer to these individuals as **success seekers**, as opposed to **failure avoiders** who have a low expectancy for success and a high fear of failure.

Failure avoiders often set low expectations and then use **defensive pessimism** (Covington, 1992) to shield themselves from experiencing failure. These individuals are motivated, but they are not happy campers. They are the students who say, "I've *got* to go to class, it's a *requirement*," rather than "I *get* to go to class, it's an *opportunity*." Please note that success-seeking and failure-avoiding are dispositional *states* and not *traits* (Safety Objective 4). Numerous situations in the workplace—from interpersonal conversations and accountability systems to management styles—can affect an unpleasant failure-avoiding disposition as depicted in Figure 6.3.

FIGURE 6.3 Working to avoid failure is not fun.

For example, how does your organization keep score with regard to safety excellence? Is it all about the total recordable injury rate (TRIR) and the number of Occupational Safety and Health Administration (OSHA) violations, or do supervisor-led work teams discuss safety-related achievements like (a) the number of hazards removed and close calls analyzed, (b) the percentage of employees who have delivered and received BBS coaching, and (c) the frequency of supportive and corrective behavioral feedback conversations delivered per week?

Indeed, many factors determine whether a work culture promotes a success-seeking or a failure-avoiding mindset regarding safety. Consider the value of determining those factors in your work culture that influence safety-related perceptions of success-seeking versus failure-avoiding. Frank and open interpersonal conversation can make such an assessment possible, but only when the next coaching lesson is followed consistently.

COMMUNICATE WITH EMPATHY

"Seek first to understand before being understood." That profound quotation from Stephen R. Covey (1989) reflects a most important concept to understand and practice to achieve a level of interpersonal discourse most likely to benefit human wellbeing. Figure 6.4 illustrates the need to perceive circumstances from the other person's perspective. Whether the topic is empathic coaching, **empathic listening**, empathic discipline, or empathic leadership, the focus is on the other person's feelings, needs, and/or perceptions (Rogers, 1961). When we begin our conversations with this mindset, we can customize

FIGURE 6.4 Mom shows a lack of empathy.

our coaching to fit the other person's perspective and be most successful at getting our point across.

When observing another individual's behavior, it is critical to try and view the context and circumstances from that person's perspective. When you listen to explanations or excuses for at-risk behavior, try to see yourself in the same situation. Imagine what defense mechanism(s) you might use to protect your ego or self-esteem. And when considering an action plan for improvement, you should attempt to view various alternatives through the eyes of the other person.

When people maintain the mindset that there is more than one side to every story, they might find another person's perspective to be very different than their own. In addition, when individuals listen with empathy to the rationale (or excuses) for a contrary opinion or behavior, they might gain appreciation for the diversity displayed and enhance mutual respect and an AC4P perspective.

Make It One-on-One

"Praise publicly and reprimand privately." Does that popular slogan sound like a good commonsense? Don't most people want to be praised publicly most of the time? Not necessarily, because some people feel embarrassed when singled out in front of a group, as depicted in Figure 6.5. Part of that

FIGURE 6.5 Public recognition can do more harm than good.

embarrassment could be fear of subsequent harassment by peers. In fact, Scott actually experienced the situation depicted in Figure 6.5. The teacher of his third-grade class asked him to stand for public recognition of turning in the best homework assignment. Later during recess, three boys from the class beat him up in the playground.

Certainly, people who deliver public recognition believe the experience will be special and positive for the recipient of the praise. In this case, the deliverer of public praise would probably prefer to receive such recognition in a public setting. This is an example of the Golden Rule: "Treat others the way you want to be treated." However, the illustration suggests that some individuals might dislike receiving public recognition from teachers, coaches, work supervisors, or public officials. This situation discredits the Golden Rule, at least to some extent.

The Platinum Rule

Better than the Golden Rule is the Platinum Rule: "Treat others the way *they* want to be treated" (Alessandra & O'Connor, 1998). Before administering a particular "treatment" or intervention (e.g., a recognition ceremony, intervention technique, or disciplinary policy), you should solicit suggestions and opinions from those who could be affected by the intervention (Safety Directive 35).

Considering another person's perspective with empathy enhances the perception of personal choice (Safety Directive 14). In order to treat others the way they want to be treated, you need to solicit their opinion, or perhaps give them a choice between alternative interventions, policies, or behavior management techniques. Implementing the Platinum Rule facilitates the perception of choice among those to be "treated." This fuels self-motivation. As W. Edwards Deming (1991) taught us years ago, "People support that which they helped to create." Solicit intervention suggestions from the target individual(s) and you will enhance intervention relevance and impact.

But the Platinum Rule does have its limits, especially when considering OHS. Safe operating procedures (SOPs) need to be followed on the job, on the road, and at home. Without required SOPs, many workers would take at-risk shortcuts, and some drivers would choose vehicle speeds way over the posted speed limit. The empathic listener must strategically disallow those behavioral options that are illegal and antisocial, or put the speaker at risk of harming him/herself and/or others.

Conclusion

Safety Objective 6 reviewed some critical research-based principles from psychological science regarding the implementation of interpersonal coaching for safety success. First, behavior-based coaching was defined with the letters of COACH—**C**are, **O**bserve, **A**nalyze, **C**ommunicate, and **H**elp. Then a most important but often overlooked component of BBS was explained—interpersonal behavioral feedback.

While supportive feedback should be directive, corrective feedback needs to be nondirective and offered with an empathic mindset. The delivery of more supportive than corrective feedback for safety-related behavior can contribute to a success-seeking safety culture and increase employees' optimism, empowerment, and self-motivation to keep each other injury-free.

However, it is neither easy nor efficient to achieve an empathic level of awareness and appreciation. Before we can listen actively and empathically to the communication from another individual, we need to minimize those reactive filters that bias our conversations. Through empathic listening, we truly prepare ourselves to coach others on behalf of their wellbeing.

DISCUSSION QUESTIONS

1. What specific work situations could benefit from AC4P safety coaching between line workers?

2. What factors hold employees back from giving each other behavior-based feedback regarding safety-related behavior?

3. How might obstacles to coworkers giving each other safety-related feedback be overcome at your workplace?

4. Please discuss a personal experience during which you *received* behavior-based feedback from a colleague regarding your safe or at-risk behavior. How effective or ineffective was that feedback, and why?

5. Please discuss a personal experience in which you *delivered* behavior-focused feedback to another individual regarding his/her safe or at-risk behavior. Was your feedback supportive, corrective, or both?

6. Please discuss one or more situations when you find it particularly difficult or awkward to give behavior-based feedback to a coworker regarding safety-related behavior.

7. How could you improve your typical feedback communications with coworkers?

8. What holds you back from having more effective AC4P feedback conversations with coworkers about their at-risk behavior?

9. Please describe a disadvantage of the Golden Rule—"Treat others the way *you* want to be treated."

10. The Platinum Rule—"Treat others the way *they* want to be treated"—incorporates empathy. Is there ever a disadvantage of the Platinum Rule? Please explain.

11. What factors, if any, influence a failure-avoiding mindset with regard to OHS in your work culture?

12. How could a success-seeking mindset or perception be promoted and supported in your work culture?

REFERENCES

Alessandra, T., & O'Connor, M. S. (1998). *The Platinum Rule: Discover the four business personalities and how they can lead you to success.* New York: Warner Business Books.

Carter, D., Flannery, N., Coiley, P., Warren, K., & Geller, E. S. (2017, August). *How coaches affect athletes: Linking behaviors to perceptions.* Poster presented at the annual conference of the American Psychological Association, Washington, D.C.

Chance, P. (2008). *The teacher's craft: The 10 essential skills of effective teaching.* Long Grove, IL: Waveland Press, Inc.

Covey, S. R. (1989). *The seven habits of highly effective people: Restoring the character ethic.* New York: Simon and Schuster, Inc.

Covington, M. V. (1992). *Making the grade: A self-worth perspective on motivation and school reform.* New York: Cambridge University.

Geller, E. S., & Geller, K. S. (2017). *Actively caring for people's safety: Cultivating a brother's/sister's keeper culture.* Park Ridge, IL: The American Society of Safety Professionals.

Geller, E. S., Perdue, S., & French, A. (2004). Behavior-based safety coaching: Ten guidelines for successful application. *Professional Safety, 49*(7), 42–49.

Reed, D., et al. (2016). Actively caring for higher education. In E. S. Geller (Ed.). *Applied psychology: Actively caring for people* (pp. 563–593). New York: Cambridge University Press.

Rogers, C. R. (1961). *On becoming a person: A therapist's view of psychotherapy.* Boston, MA: Houghton Mifflin.

Seligman, M. E. P. (1991). *Learned optimism.* New York: Doubleday/Currency.

Thorndike, E. L. (1931). *Human learning.* Cambridge, MA: MIT Press.

7 Safety Objective 7

Achieve Synergy with Interdependence and Systems Thinking

Safety Directive 37: Apply the power of synergy.
Safety Directive 38: Optimize synergy with diversity.
Safety Directive 39: Distinguish between trusting ability vs. intention.
Safety Directive 40: Practice the seven C's of trust-building.
Safety Directive 41: Become a systems thinker.
Safety Directive 42: Promote and support an AC4P culture.

Safety leaders promote diversity, interdependency, and **interpersonal trust**; they realize these qualities of an organizational culture are critical for achieving and maintaining an injury-free workplace. Safety Objective 7 explains these qualities of an AC4P work culture with safety-related definitions. We begin by considering an optimizing mission of safety leadership—the achievement of synergy.

THE POWER OF SYNERGY

Google defines synergy as "the interaction or cooperation of two or more organizations, substances, or other agents to produce a combined effect greater than the sum of their separate effects." A beneficial synergistic effect occurs when *divergent* ideas or perspectives combine to derive an action plan more effective than one attainable from individuals working alone and independently. Synergy is possible when a mindset of win/lose independency shifts to win/win interdependency (Safety Objective 7). Note the modifier used when defining the input needed to obtain a synergistic effect—divergent.

If everyone on a problem-solving team has similar expertise, issue-related viewpoints, and educational backgrounds, deriving and agreeing on a relevant action plan will be relatively quick, but then the outcome reflects input from only the sum of similar parts. A synergistic action plan occurs when the parts are dissimilar and interact to produce an innovative solution that might be a

DOI: 10.1201/9781003758822-8

compromise between divergent perspectives. Reaching a consensus or compromise between participants with disparate opinions does not happen efficiently, but it is likely that the action plan derived will be most effective.

Perhaps a discussion of politics is illustrative of maximum disparity and difficulty at achieving consensus and synergy, even though the human welfare objectives for Democrats and Republicans are more similar than dissimilar, as illustrated in Figure 7.1. That is why people are frequently advised: "Don't discuss politics." It is apparently assumed that people cannot be open-minded regarding politically charged issues.

FIGURE 7.1 Win/lose stifles synergy.

Discussing politics will not likely change opinions, but it can enable an understanding and appreciation of divergent viewpoints. With openness and empathy, such communication could actually lead to a useful teaching/learning experience.

Diversity

A high-performance safety team includes people with knowledge, skills, and abilities that complement each other's competencies. Diversity benefits group decision-making and action planning, even though obtaining a consensus from a diverse group is relatively cumbersome and time-consuming. Safety leaders realize that the synergistic results of individual contributions increase as a function of the diversity of ideas, opinions, and mindsets among the participants.

All of this is easier said than done. It is more efficient and convenient to just sit back and go along, giving up individual identity or a sense of personal responsibility. This is analogous to the diffusion of responsibility phenomenon used to explain the **bystander apathy effect**—the observation that the more people present at an emergency, the less likely will any one individual help (Latané & Darley, 1968, 1970). As depicted in Figure 7.2, minimal diversity usually results in most efficient deliberation and decision-making, but the resultant action plan will most likely be suboptimal and/or unjust.

FIGURE 7.2 Diversity benefits decision-making.

Making diversity work well to obtain optimal synergy is a key challenge for a facilitator or leader of a safety team. Innovative ideas and solutions to problems are most likely to emerge from combinations of divergent competencies, experiences, paradigms, and yes—biases. This level of interactive dialog and interpersonal learning leads to optimal synergy. Perspectives and life experiences mix and match to activate a creative birth and death of ideas. Perspectives and experiences evolve into a whole of greater understanding and innovation (Geller, 2025).

The synergistic whole is greater than the sum of its diverse parts because there is synergy between the parts. The parts contribute to the whole and manifest its essence, its purpose, and its function. Thus, the whole has a fundamental relationship with its parts. Indeed, the whole safety committee, group, or team is the reason for the existence of its parts. For this process to work most effectively, however, the parts—the people—must maintain a mutual interpersonal trust for each other's ability, intentions, and potential contributions.

Cultivating Interpersonal Trust

Effective supervisors, safety professionals, and members of a synergistic safety team are trustworthy, and they engender trust among others. What does that mean? You could be confident a person means well, but you might doubt his/her ability to complete the intended task. In this case, you trust the individual's intentions, but you lack confidence in the capability of this person to make good on his/her promises. "She is well intended, but she doesn't have what it takes to make it happen."

In contrast, people can have faith in the ability of others, but mistrust the intentions of these individuals. The followers might believe their leader has the intellect and the skills to make a difference, but they might not be convinced their welfare is considered in this person's decision-making. "He has the talent and support system to make this happen, but he's looking out more for his own best interests than for ours."

The following seven C-words represent strategies for building interpersonal trust in a person's intentions and ability: Communication, Caring, Candor, Consistency, Commitment, Consensus, and Character (Geller, 2002). The trust-building utility of these C-words is reflected by their definitions and common usage. Let's consider briefly the definition of these trust-determining words as they relate to benefiting OHS.

Communication. What people say and how they say it influence our trust in both their capability and their intentions. An individual's expertise is displayed by his/her spoken or written words and by the confidence and credibility linked to those words. However, the way something is said, including intonation, pace, facial expressions, hand gestures, and overall posture, can have greater impact than what is actually said. In fact, it is likely many readers have experienced a change in their personal feelings of trust toward another person as a result of how that individual communicated information. As illustrated in Figure 7.3, a lack of interpersonal trust can be communicated without words.

Caring. When you take the time to listen to another person's perspective, you send a most important message—you care. When people believe you sincerely care about them, they will care about what you tell them. They trust that you will look out for them when applying your knowledge, skills, and abilities. They trust your intentions because they believe you care.

Candor. We trust people who are frank and direct with us. They get right to the point, whether asking for a favor or giving us feedback about our behavior. When these individuals don't know an answer to our questions, they don't ignore us or blather ambiguously about divergent possibilities. They tell us outright when they do not know something, and they tell us they will get back to us later. When they get back to us soon with an answer, our trust in both their intention and ability increases.

FIGURE 7.3 Behavior can reveal trust or mistrust.

The second definition of Candor is "freedom from prejudice" (*The American Heritage Dictionary*, 1982, p. 233). This reflects another important aspect of trust-building. When a person's interactions suggest prejudice or the tendency to evaluate or judge another person because of a stereotype or preconceived notion about group characteristics, you have reason to mistrust that individual. You are rightfully suspicious about that person's ability and intentions to evaluate and treat people fairly.

Consistency. Perhaps the quickest way to destroy interpersonal trust is to not follow through consistently with interpersonal agreements. A promise is essentially a behavior-consequence contingency. It specifies that a certain consequence will follow a certain behavior. Whether the consequence is positive or negative, trust decreases when the behavior is not rewarded or penalized as promised.

Commitment. People who are dependable and reliable are not only showing consistency, they are demonstrating commitment. When you follow through on a promise or a pledge to do something, you tell others they can count on you. You can be trusted to do what you say you will do. Making and honoring a commitment builds trust in both intention and ability.

Consensus. When a group of employees reaches consensus about something, everyone agrees on a decision or a course of action and is willing to support it. Safety leaders who develop consensus among their followers build interpersonal trust. Consensus-building is the opposite of top-down decision-making. It is also not the same as negotiating or calling for a vote and letting the majority win.

Consensus-building is certainly not easy. There is no quick fix to arriving at a consensus. Plenty of interpersonal communication is required, including straightforward opinion sharing, intense discussion, emotional debate, active listening, careful evaluation, methodological organization, and systematic prioritizing. However, on important matters, the outcome is well worth the investment. When a leader develops a solution or process that every potential participant can get behind and champion, the degree of interpersonal trust needed for total involvement has been cultivated.

Character. Consider that people with character are honest, ethical, principled, and worthy of another person's trust. They have the courage to admit their mistakes with a sincere apology and a request for forgiveness. These trustworthy individuals avoid talking about deficiencies in others. They realize that publicly criticizing or demeaning other individuals leads to interpersonal suspicion and mistrust. They commit to a policy of "No Backstabbing." They always talk about others as if those other people can hear them.

SYSTEMS THINKING

"Don't blame people for problems caused by the system" (Deming, 1991). That profound quotation from W. Edwards Deming reminds us to focus our efforts on optimizing the system (Deming, 1986). When assessing, analyzing, or solving a problem, systems thinkers attempt to see the big picture. They diagnose close calls and injuries with care, and they certainly do not look for a root cause (Safety Objective 4). They take a broad view of the situation and consider the dynamic and reciprocal interaction among the many interdependent situational and dispositional factors that influence unfortunate and unintentional incidents.

Take the common behavior of driving a vehicle, and consider how an independent versus an interdependent focus impacts behavior and attitude. Unlike the systems thinker in Figure 7.4 who perceives the bigger picture, too many drivers demonstrate an individualistic win/lose mindset rather than the win/

win cooperative perspective needed to optimize the transportation system. They dart back and forth between lanes, often without signaling, just to shave a few seconds off their trip time. Given a line of vehicles in the left lane, these individualistic drivers pass the vehicles on the right and then use their left-turn signal to squeeze in front of the line.

FIGURE 7.4 Systems thinkers see the big picture.

How about those slow drivers clogging up the left lane, perhaps because they will eventually make a left-hand turn several miles ahead? Or what about those drivers who use a cell phone while driving and take extra time at intersections because they are reading or sending a text message? Those drivers are also displaying an individualistic win/lose perspective—perhaps unconsciously or unintentionally–which makes the transportation system less efficient and effective.

Few of those drivers probably consider their impact on the overall system—a system that works best when everyone follows the same rules, norms, and courtesies of the road. Perhaps the driver in the left lane who allows space for the self-serving driver who passed the line of vehicles on the right sees the bigger picture and realizes how a simple act of considerate AC4P driving helps to optimize the system.

Conclusion

"Culture" is the hot buzzword for improving OHS these days, although the concept of organizational culture or climate has been a frequent topic at safety conferences for decades (Safety Objective 10). In fact, Scott introduced the need to cultivate an interdependent AC4P culture in his first textbook on the human dynamics of OHS in 1996 (Geller, 1996). He emphasized that the spirit of win/win interdependence is vital to meet the challenge of attaining and sustaining an injury-free workplace.

Safety Objective 7 identified critical characteristics of an AC4P culture and suggested some strategies for obtaining an interdependent culture of people routinely performing AC4P behavior on behalf of the safety and health of their coworkers.

The first safety directive reflects the power of synergy—interpersonal collaboration that results in an optimal outcome beyond what could be achieved by participating individuals operating independently. The next safety directive explains why such optimization requires diversity, interpersonal trust, and systems thinking. Then we introduced seven C-words that suggest distinct directives for building interpersonal trust (Safety Directive 40).

Communicating these trust-building guidelines to others in a *candid* and *caring* way opens up the kind of dialogue that can start people on a journey of interpersonal trust-building. Next, people need to give each other *consistent* and *candid* feedback about how their behavior relates to these trust-building principles.

With *character* and *commitment*, people need to recognize each other for participating in cultivating interpersonal trust and offer corrective feedback when there is room for improvement (Safety Objective 6). Recipients of such constructive feedback must accept legitimate correction with *caring* gratitude and a *commitment* to improve.

Feedback recipients need to show *character*. They thank observers for their AC4P feedback, even when the *communication* is not all positive and/or is not delivered well. They might offer feedback on how to make the feedback they received more useful. Dialogue like that is necessary to build *consensus* and sustain a journey of continuous learning and trust-building.

Systems thinkers realize how applications of these seven trust-building words enable the cultivation of an interdependent AC4P work culture. Systems thinkers also diagnose incidents with care and do not look for a root cause. They obtain a broad picture of the situation and consider the dynamic and reciprocal interaction among various environmental, behavioral, and dispositional factors.

Systems thinkers look for and attempt to improve the cultural factors that contributed to unfortunate and unintentional incidents—from close calls and minor injuries to serious injuries and fatalities.

DISCUSSION QUESTIONS

1. Please explain the concept of synergy with regard to improving OHS at your workplace.

2. How can diversity improve safety-related decision-making and incident analysis at your workplace?

3. Is the distinction between trusting *intention* versus *ability* relevant for anyone at your workplace? Please explain.

4. How might a leader or a group facilitator promote consensus-building and why is that important?

5. What does it mean to be candid when communicating about safety-related issues, and why is that important?

6. What are some OHS-related disadvantages of negative gossip and backstabbing?

7. How does Dr. Deming's profound quotation—"Don't blame people for problems caused by the system"—relate to OHS?

8. What are some critical characteristics of an AC4P culture, and how could some of these qualities be achieved at your workplace?

REFERENCES

American Heritage Dictionary. (1982). Second College Edition. New York: Houghton Mifflin Company.

Deming, W. E. (1986). *Out of the crisis*. Cambridge: Center for Advanced Engineering Study, Massachusetts Institute of Technology.

Deming, W. E. (1991, May). *Quality, productivity, and competitive position*. Four-day workshop presented in Cincinnati, Ohio by Quality Enhancement Seminars, Inc., Los Angeles.

Geller, E. S. (1996). *The psychology of safety: How to improve behaviors and attitudes on the job*. Boca Raton, FL: CRC Press.

Geller, E. S. (2002). *The participation factor: How to increase involvement in occupational safety*. Des Plaines, IL: American Society of Safety Engineers.

Geller, E. S. (2025). Diversity, equity, and inclusion: Behavioral science interpretations for optimal outcomes. *Behavior and Social Issues, 34*, 281–292. DOI:10.1007/s42822-025-00217-5

Latané, B., & Darley, J. M. (1968). Group inhibition of bystander intervention. *Journal of Personality and Social Psychology, 10,* 215–221.

Latané, B., & Darley, J. M. (1970). *The unresponsive bystander: Why doesn't he help?* New York: Appleton-Century-Crofts.

8 Safety Objective 8
Apply Humanism to Make BBS More Effective

Safety Directive 43: Treat people as individuals.
Safety Directive 44: Listen with empathy.
Safety Directive 45: Cultivate self-accountability.
Safety Directive 46: Pay attention to intention.
Safety Directive 47: Understand motivation with Maslow's Hierarchy of Needs.
Safety Directive 48: Teach and experience self-transcendence.

Safety Objective 1 illustrated the complexity of the human side of safety by reviewing seven dimensions of people—Behavior, Attitude, Sensation, Imagery, Cognition, Interpersonal, and Drugs (BASIC ID). We recommended addressing the human dynamics of safety by starting with behavior and the mission to "Act people into constructive safety-related attitudes, perceptions, cognitions, images, and interpersonal support."

Although readers could interpret that recommendation as a promo for BBS, that quotation implicates a critical role of human dynamics beyond observable behavior. Safety Objective 8 provides an evidence-based framework for our recommendation, as well as some practical intervention strategies.

HUMANISM VS. BEHAVIORISM

Readers who have experienced an introductory psychology course undoubtedly have heard of both **behaviorism**, made popular by B.F. Skinner (Skinner, 1953, 1974), and **humanism**, developed by Carl Rogers (1961, 1980). It is likely that readers heard the introductions to these philosophies and approaches to psychotherapy as opposing, even competing, perspectives. We hope to convince you otherwise—that these intervention approaches can be mutually complementary and supportive.

Behaviorists treat only the behavior of their clients, essentially applying positive consequences for desirable behavior and removing positive consequences for undesirable behavior (Safety Objective 2).

DOI: 10.1201/9781003758822-9

In contrast, humanists target and evaluate people's intentions, and focus on discovering a client's personal perceptions, motives, and self-concept. Behavioral therapists are selective when providing positive consequences (e.g., positive recognition, supportive feedback, and behavior-based rewards); humanistic therapists give positive regard unconditionally.

The humanist's clinical approach is nondirective. The therapist does more listening than directing. In contrast, behavioral therapists are directive. They define behavioral consequences to change in order to increase the frequency of desirable behavior and/or decrease occurrences of undesirable behavior.

We could point out other distinctions between behaviorism and humanism, but these few distinctions explain why Scott has received negative body language whenever his keynote addresses at professional conferences for applied behavior analysts (or applied behavioral scientists) propose "humanistic behaviorism" as an intervention approach. He has stifled some negative vibes by informing the audience that B.F. Skinner was honored with "Humanist of the Year" in 1972 and said the following: "Behaviorism makes it possible to achieve the goals of humanism more effectively" (Skinner, 1978, pp. 9–10).

A BENEFICIAL COMPROMISE

We reverse Skinner's quotation: "*Humanism* makes it possible to achieve the goals of *behaviorism* more effectively." For safety leaders, we propose "The application of select principles from humanism can optimize the injury prevention impact of the essential intervention process of BBS—peer-to-peer observation and feedback." In fact, many readers who have applied an effective BBS coaching process have likely used some of these principles, perhaps without realizing they were practicing humanism.

Let's consider five critical principles of humanistic therapy as potential positive support for BBS, thereby justifying the label—**humanistic behaviorism.**

1. Treat People as Individuals.

   Humanistic psychologists adhere to a philosophy of phenomenology, meaning everyone experiences his/her surroundings differently. As depicted simply in Figure 8.1, people view their life space from their own personal vantage point; it is impossible and inappropriate to interpret another person's experience. The only way to understand an ongoing interaction with another person and his/her current circumstances is to ask questions and listen actively and openly without personal judgment or interpretation.

FIGURE 8.1 Perception is personal.

It is natural to relate one's own experiences to another person's story and draw parallels for mutual understanding and advice. However, humanists actively avoid such personal interpretation, realizing that their own idiosyncratic perceptions could bias an analysis and/or appreciation of the storyteller's perspective.

Note how this idiographic or phenomenological approach contradicts many everyday attempts to explain the behavior of others, as when analyzing the contributing factors to a workplace injury (e.g., the Fundamental Attribution Error of Safety Objective 4). Moreover, the common therapy or counseling technique of interpreting an individual's emotional conflict or person-state with reference to a particular theory or concept is also opposite to this humanistic approach. Generalizing a theory or construct across individuals is considered nomothetic—opposite to the idiographic or phenomenological approach advocated by humanists.

Figure 8.2 illustrates a most serious issue with generalizing and not treating people as individuals. We are talking about stereotyping, which occurs when evaluations are influenced by a person's status in a particular identifiable group, such as "supervisor," "safety professional," "student," "patient," "line worker," "union representative," "athlete," or "homeless person." Each label activates a particular image and various characteristics. Then, the general label we give people influences how we view them, judge them, and react to their

FIGURE 8.2 Discrimination is good; stereotyping is bad.

conversations with us (Judd, Ryan, & Park, 1991). When we prejudge people on such generalizations, we are practicing prejudice or bigotry.

Efforts to combat prejudice focus on teaching people that they should consider everyone equal and stop categorizing them. Stop "discriminating." Ellen Langer (1989) advocates another approach to discrimination. Categorizing people and things according to discernible characteristics is a natural learning process; it is how we come to know and understand people and their surroundings.

The key to reducing stereotyping and prejudice is to make more, not fewer, distinctions between people, as humanists advocate. When people become more attentive to the numerous differences between individuals and understand how those differences vary according to the environmental or interpersonal context, it becomes increasingly difficult to generalize across individuals and put them into generic categories. Notice how recognizing and appreciating distinctions between people optimizes the achievement of synergy (Safety Objective 7).

2. Listen with Empathy.

It is neither easy nor commonplace to adhere to the idiographic approach of humanism, especially in our fast-paced society of "getting things done as efficiently as possible." Who has the time to listen actively to every individual's personal story before making a decision? Combine this time-urgency mindset with contemporary digital communication referenced in Figure 8.3, and this humanistic approach to addressing and improving human wellbeing becomes even more challenging.

FIGURE 8.3 Digital communication is limiting.

Humanists set a high bar for understanding and helping others, especially among individuals in a work setting with limited time and opportunity for the level of interpersonal conversation alluded to here. Naturally, a high level of humanistic relationship-building is essential and common among family members, and perhaps with select individuals in a work setting. Indeed, any attempt to approximate this humanistic approach with colleagues on a work team would certainly benefit everyone's safety and health.

Listening to another person's opinion or perspective with empathy before providing advice or direction reflects Stephen Covey's fifth habit of highly effective people—"Seek first to understand, then to be understood" (Covey, 1989). As most readers know, empathic listening is more than feeling sympathy and expressing concern for another person's plight. We truly empathize when we identify with another person's situation and attempt to understand what it is like to be in the other person's "steel-toed shoes" (Safety Directive 35).

The critical behavior checklist (CBC) used in BBS to observe and record safe and at-risk behavior is not handed down by OSHA, management, or a safety professional. No, the CBC is developed by a team of workers on a particular job, and when their job and/or the environmental setting changes, these employees modify this CBC following open and candid interpersonal communication with their work team. That is empathy in action.

Plus, when humanistic coworkers, safety professionals, and supervisors give an employee corrective feedback for at-risk behavior observed, they practice empathy (Safety Directive 35). They do not begin with behavior change directives. Instead, they ask questions to understand the rationale for the at-risk behavior and to learn if features of the situation could be altered to facilitate occurrences of the safe alternative.

3. Cultivate Self-Accountability.

   Rather than telling an employee what safe behavior should replace an observed at-risk behavior, a humanistic observer asks the worker what s/he could do to reduce the probability of an injury and set the safe example for others. Similarly, after identifying a work problem, humanistic supervisors do not specify a resolution. Instead, they challenge the relevant employees to discuss possible solutions and propose an action plan. As every reader has experienced, ownership and self-accountability happen when individuals perceive some personal autonomy and receive respect and appreciation for their competence to collaborate with peers at solving an issue (Safety Objective 3).

This commonsense strategy for facilitating self-accountability or self-directed behavior is founded on self-determination theory derived from humanistic theory and research (Deci, 1975; Deci & Flaste, 1995). In particular, perceptions of autonomy (or choice), competence, and relatedness (or interdependence) enhance self-accountability. Figure 8.4 illustrates the importance of self-accountability or self-motivation. The worker is alone with no one but himself and his self-talk

FIGURE 8.4 Injury-free requires self-accountability.

to activate and support safe behavior. Without self-accountability for safety, this worker will likely choose to not wear his hardhat—the more comfortable and convenient choice.

4. Pay Attention to Intention

Humanists focus more on people's intentions than their behaviors. Suppose you observed two male college students pushing and shoving each other in a university parking lot after a football game. What label would you associate with that behavior? A behaviorist would likely call the altercation "aggression," because the interpersonal behavior typifies physical conflict. The humanist, however, would interpret that observation differently, claiming one should not label such behavior "aggression" without considering the intentions of both participants.

What if the behaviors observed reflected a friendly physical exchange, as in "horseplay"? Suppose the congenial pushing and shoving changes to unfriendly physical conflict, perhaps because a "friendly" push caused some pain or discomfort to the recipient. Both Krista and Scott have experienced such a shift in the spirit of an analogous encounter, and we believe many readers have been in a comparable situation. The interpersonal behavior is similar, but intentions have changed. While the behaviorist would maintain the label of "aggression," the humanist would account for intentions and change the behavioral label from "horseplay" to "aggression." How does this apply to safety?

For Safety Objective 6, we introduced the observation-and-feedback process of BBS and the application of a CBC to record observations of a coworker's safe and at-risk behavior. After completing a CBC, the observer should conduct a brief behavioral feedback session with the worker observed and review the CBC results—the frequency of safe and at-risk behavior. As indicated in our discussion of Safety Objective 6, this critical interpersonal feedback component of a BBS observation-and-feedback session is often overlooked or handled incompletely. Why are peer-to-peer discussions of the CBC results unpopular? Answer: These conversations take time and can feel awkward, especially if the observer had recorded one or more at-risk behaviors.

How do you tell a coworker s/he had been working unsafely and increasing the probability of a serious injury? For Safety Objective 6, we explained the nondirective approach to giving corrective feedback, but at this point we want readers to consider the utility of discussing "intentions" as a way to increase the occurrence of informative interpersonal BBS feedback sessions.

Figure 8.5 illustrates a flaw in the typical approach to observing and explaining the behavior of others. Observers interpret the behavior they perceive from their own experience. But only the performers themselves can provide an accurate explanation for their behavior—their intentions. Could the interpersonal conversation about safe and at-risk behavior be more influential than only showing which columns the observer checked on the CBC of a BBS observation-and-feedback process?

FIGURE 8.5 Personal experience biases interpretation.

Consider the value of discussing a person's intentions for performing safe or at-risk behavior. Of course, the intention or rationale for working safely is obvious—to prevent an injury. Still, verbalizing this intention could actually increase the frequency of safe behavior, especially among lone workers—when no one is watching. Such overt verbal behavior can influence relevant cognition or self-talk, perhaps shifting habitual behavior (unconscious competence) to mindful fluency (conscious competence).

The advantage of focusing on intention is greatest for conversations about at-risk behavior. An open discussion of such intentions could reveal injury prevention factors (e.g., management and peer influence, environmental context, and reward/penalty contingencies), as well as dispositional determinants (e.g., attitude, cognition, or fatigue) that could have influenced the occurrence of an at-risk behavior. "I intended to follow the SOP, but…." Some of the factors revealed in a feedback conversation about intention could decrease the probability of an error or an injury, if modified accordingly.

It is likely the quantity and quality of interpersonal conversations about CBC results will increase if the focus is not on "observing and recording occurrences of safe and at-risk behavior," but on "increasing safety-related mindfulness and discovering factors that might influence some workers to perform at-risk behavior." Of course, the CBC is completed and recorded as usual, but accomplished openly to reveal situational and dispositional determinants of safe and at-risk behavior by discussing "intentions."

Question: Could this mindset or expectancy enhance the acceptance, applicability, and accuracy of CBC results and increase the frequency of beneficial post-observation conversations? We think so.

5. Appreciate Maslow's Hierarchy of Needs.

Every reader who has taken a course on human motivation has heard about this approach to understanding variations in human motives, created by humanist Abraham Maslow (1970). Categories of needs are arranged hierarchically, and it is presumed that we do not attempt to satisfy needs at one level until our needs at the lower levels are satisfied to some degree.

First, we are motivated to fulfill our physiological needs—basic survival requirements for food, water, shelter, and sleep. After these needs are under control, we are motivated by the desire to feel secure and safe from potential dangers. Next, we have our social acceptance needs—to have friends and feel a sense of belonging. When these needs are gratified, our concern focuses on self-esteem–earning self-respect and feeling worthwhile.

After enjoying a boost in self-esteem, we presumably reach the top of this need hierarchy–**self-actualization** or the realization of achieving one's full potential. While many have learned that self-actualization is atop this need hierarchy, Maslow revised his hierarchy near the end of his life by placing another ultimate achievement at the top–self-transcendence (Maslow, 1971). As depicted in Figure 8.6, we are the best we can be when we reach beyond our own self-interests and contribute to the needs of others. Isn't that exactly what safety leaders do on a daily basis? They intervene whenever possible to keep others safe from personal injury.

How satisfying to realize that you reach the top of Maslow's Hierarchy of Needs every time you act on behalf of another person's health and/or safety. And doing this helps to satisfy your lower-level needs that never get completely satiated–social acceptance, self-esteem, and self-actualization. Consider this: Achieving

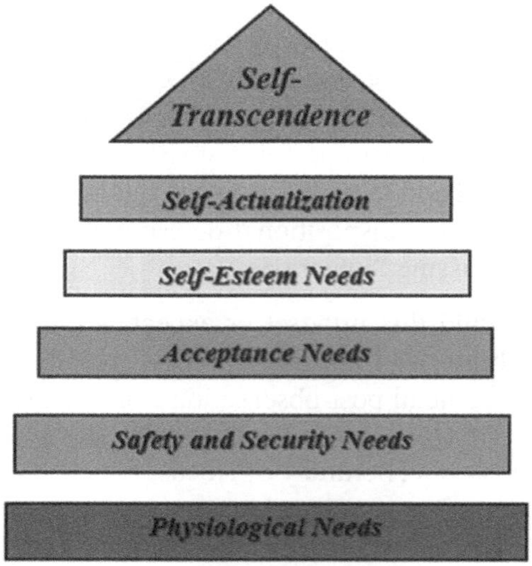

FIGURE 8.6 The top of Maslow's Hierarchy is not self-serving.

self-transcendence is the ultimate outcome and is self-reinforcing because it naturally satisfies other higher-level needs. The more people who teach and experience the reciprocal positive effects of self-transcendent behavior, the closer we come to an AC4P culture.

Conclusion

Consulting firms have used a variety of labels to market programs designed to address the human dynamics of occupational safety, and those labels influence the nature of the intervention approach. For example, Scott used the term "behavior-based safety" (BBS) in 1979 when developing and evaluating intervention strategies to increase the use of vehicle safety belts among the employees of Ford Motor Company (Geller, 1979).

That safety intervention approach was among the first to put employees in control of implementing and evaluating techniques to increase occurrences of safe behavior and reduce the frequency of at-risk behavior. One of those techniques was peer-to-peer behavior-focused observation and feedback (Safety Objective 6). After several consulting firms marketed this approach for preventing workplace injuries on a large scale, BBS quickly became a worldwide intervention strategy for OHS.

After noting that most applications of BBS focused exclusively on employees' behavior, Scott made the case for another label—"people-based safety" or

PBS (Geller, 2005, 2008). PBS implicates consideration of more than behavior when designing and implementing a procedure to address the human dynamics of occupational safety. Several organizations, as well as the worldwide consulting and training practices of Safety Performance Solutions, Inc., called their observation-and-feedback process PBS rather than BBS to emphasize that their interventions target and influence more than employees' behavior.

Safety Objective 8 offers yet another label for a similar injury prevention process—humanistic behaviorism or actively (*behavior*) caring (*humanism*) for people (AC4P). To demonstrate a reasonable connection between humanism and behaviorism, we identified five basic assumptions of humanism to consider when addressing the human side of safety, particularly the delivery of behavior-based feedback (Safety Objective 6). Incorporating these assumptions into a peer-to-peer observation-and-feedback process will not alter the development and application of a CBC to observe and record occurrences of safe and at-risk behavior.

Adopting certain assumptions of humanism—particularly empathic listening, attention to intention, and the determinants of self-accountability or self-motivation (Safety Objective 3)—would benefit the crucial interpersonal feedback communication following a BBS observation session. In fact, a humanistic mindset increases both the quantity and the quality of post-observation discussions of safe versus at-risk behavior. Such interpersonal conversations can reveal invaluable information for understanding occurrences of at-risk behavior and for modifying environmental and management-system factors relevant for decreasing incidents of human error and workplace injury.

DISCUSSION QUESTIONS

1. What are some critical distinctions between Carl Roger's humanism and B.F. Skinner's behaviorism?

2. What behaviors do people need to perform in order to treat others as individuals?

3. Explain the difference between discrimination and stereotyping.

4. List the advantages and limitations of digital communication (e.g., email, text messaging, LinkedIn, and Facebook).

5. What principles of humanism are relevant to inspire more employee engagement for OHS?

6. How can attention to intention benefit a behavior-based coaching process?

7. Distinguish between giving directive versus nondirective corrective feedback with a realistic workplace example.

8. Imagine you observe an employee at your workplace performing an at-risk behavior and you have an opportunity to offer some one-to-one feedback. How might you refer to both intention and behavior in that interaction?

9. How does Abraham Maslow's Hierarchy of Needs relate to B.F. Skinner's "selection by consequences"?

REFERENCES

Covey, S. R. (1989). *The seven habits of highly effective people: Restoring the character ethic* (pp. 236–260). New York: Simon & Schuster.
Deci, E. L., (1975). *Intrinsic motivation*. New York: Plenum.
Deci, E. L., & Flaste, R. (1995). *Why we do what we do: Understanding self-motivation*. New York: Penguin Book.
Geller, E. S. (1979). *Behavior-based safety to increase the use of vehicle safety belts*. Detroit, MI: Ford Motor Company.
Geller, E. S. (2005). *People-based safety: The source*. Virginia Beach, VA: Coastal Training and Technologies Corporation.
Geller, E. S. (2008). *Leading people-based safety: Enriching your culture*. Virginia Beach, VA: Coastal Training and Technologies Corporation.
Judd, C. M., Ryan, C. N., & Park, B. (1991). Accuracy in the judgment of in-group and out-group variability. *Journal of Personality and Social Psychology, 61*, 745–755.
Langer, E. J. (1989). *Mindfulness*. Reading, MA: Addison Wesley.
Maslow, A. H. (1970). *Motivation and personality* (2nd ed.). New York: Harper & Row.
Maslow, A. H. (1971). *The farther reaches of human nature*. New York: Viking Press.
Rogers, C. R. (1961). *On becoming a person: A therapist's view of psychotherapy*. Boston, MA: Houghton Mifflin.
Rogers, C. R. (1980). *A way of being*. Boston, MA: Houghton Mifflin.
Skinner, B. F. (1953). *Science and human behavior*. New York: Macmillan.
Skinner, B. F. (1974). *About behaviorism*. New York: Alfred A. Knopf.
Skinner, B. F. (1978). *Reflection on behaviorism and society* (pp. 9–10). Englewood, NJ: Prentice Hall.

9 Safety Objective 9
Nurture a Healthy OHS Mindset within Yourself and among Others

Safety Directive 49: Accept stress but stifle distress.
Safety Directive 50: Distinguish between personal control versus concern.
Safety Directive 51: Become a hopeful optimist.
Safety Directive 52: Accept Type A *behavior* but stifle Type A *emotion*.
Safety Directive 53: Apply constructive self-talk.
Safety Directive 54: Adopt a mindset of continuous teaching/learning.

Safety success is influenced by certain person-states—dispositional perspectives, beliefs, expectancies, or mindsets. Yes, people use a variety of terms to reflect individual differences that affect behavior. We refer to this human dynamic as a person-*state*, which unlike a personality *trait* can change as environmental conditions and/or contingencies change. Safety Objective 9 addresses certain person-states reflecting a mindset that can influence the development, implementation, and impact of interventions for preventing injuries in the workplace and beyond.

Safety Objective 3 defined the psychological definition of empowerment with three questions: "Can I do it?" "Will it work?" and "Is it worth it?" Answers to these questions reflect three beliefs or person-states that determine whether a person "feels" empowered, which is not necessarily the same as "being empowered." A manager might give you the responsibility and authority to take on a particular assignment, but you might not believe you can accomplish it successfully. Feeling empowered actually implicates another critical person-state—personal control. This particular disposition or mindset determines whether a person experiences motivating **stress** or debilitating **distress** when addressing a challenge.

STRESS VS. DISTRESS

When people say they feel "stressed," they usually mean they feel overwhelmed. However, this feeling or perception is not "stress"; it is distress.

DOI: 10.1201/9781003758822-10

The difference between these two person-states is the perception of personal control. The first definition of stress in *American Heritage Dictionary* (1991) is "importance, significance, or emphasis on something" (p. 490).

When we are stressed, we are focused and motivated to make something happen. We feel empowered, especially if we also have self-efficacy ("I can do it"), response efficacy ("It will work"), and outcome expectancy ("It's worth it"). The person who claims, "I work best under pressure," understands the meaning and benefit of stress. Hans Selye, the Austrian-born founder of psychological stress research said, "Complete freedom from stress is death" (Selye, 1974, p. 32).

Distress is a harmful feeling state or mindset. It is defined as "anxiety or suffering, severe strain resulting from exhaustion or an accident" (*American Heritage Dictionary*, 1991, p. 410) or "suffering of body or mind; pain, anguish; trouble, misfortune… a condition of desperate need" (*New Merriam-Webster Dictionary*, 1989, p. 224).

We experience stress when we are managing our stressors effectively; we are distressed when we are unprepared or ill-equipped to deal with a particular stressor. A mindset of relevant personal control turns distress into stress. Anything that increases a feeling of empowerment should also increase one's perception of personal control and reduce distress.

Both runners in Figure 9.1 are experiencing a stressor, as indicated by the butterflies in their stomachs. However, the butterflies are disorganized in the runner on the left, but they are all lined up and organized in the runner on

FIGURE 9.1 Stress is good; distress is bad.

the right. The runner on the left perceives limited personal control and is distressed; the runner on the right feels in control and stressed.

Consider the following sequence of cognitive (self-talk) decisions that accompany the anticipation or occurrence of an important event (e.g., a safety talk, a performance appraisal, or a certification exam): First, you decide whether the event is important. If judged important, the event is a stressor, and then you make a second cognitive appraisal: "Do I have sufficient personal control over this stressor?" If your answers to the three empowerment questions referred to above are "Yes," you experience motivating stress. Alternatively, if your appraisal of personal control is "No," the stressor results in debilitating distress.

Domain of Influence vs. Concern

Giving people more knowledge, resources, and interpersonal support should increase their perception of personal control and prevent stress from becoming distress. However, personal control is a personal perception or mindset. That *person-state* is in the eyes of the beholder. Note that we cannot really control everything, as illustrated in Figure 9.2.

Stephen Covey (1989) recommended that we distinguish between our "circle of concern" and our "circle of influence," and focus our efforts on our

FIGURE 9.2 We can't control everything.

domain of influence. It is healthy to admit there are things we are concerned about but have little influence over—from the weather and politics to various management decisions that affect our lives. When negative consequences occur beyond our domain of influence, we do not experience distress because of a lack of personal control.

Sometimes events occur within our domain of influence, but we do not get distressed because we blame situational factors beyond our control. As explained in Safety Objective 4, this misattribution of human error is termed the Self-Serving Bias (Safety Directive 23), and it can markedly influence the quantity and quality of participation in the analysis of a close call or an injury.

It is best to start an analysis of an unfortunate and unintentional incident by discussing relevant environmental factors. After a constructive discussion of potential environmental factors contributing to an injury, people are more likely to participate in a frank and open discussion of possible dispositional or behavioral factors that might have contributed to the incident.

While the Self-Serving Bias protects self-esteem and decreases the probability of experiencing distress, it can stifle attempts at personal improvement. As discussed earlier (Safety Directive 18), you can expand your domain of influence by setting goals that are empowering and SMARTS—"S" for Specific behaviors to perform, "M" for a Motivational connection to the consequences to be gained from goal attainment, "A" for setting an Attainable stretch goal, "R" for a goal that is Relevant to a personal mission or ultimate vision, "T" for a goal that is Trackable for assessing progress toward achieving the goal, and "S" for Sharing the goal with others for social support and interpersonal accountability.

An Optimistic Mindset

Feeling optimistic usually means an individual feels in control of a particular situation and experiences more stress than distress. The statement, "I expect the best," reflects an optimistic mindset. As illustrated in Figure 9.3, optimistic individuals maintain the expectation that their life events, including personal decisions and actions, will turn out well. As discussed for Safety Objective 6, most success seekers are optimistic, in contrast to failure avoiders who are more likely to be pessimistic.

Compared to the pessimist, optimists maintain a sense of humor, perceive problems or challenges in a positive light, and plan for a successful future. They focus on what they can do—their behavior—rather than on how they feel (Carver, Scheier, & Weintraub, 1989). Optimists handle their stressors

FIGURE 9.3 Optimists experience less distress.

constructively and experience more positive stress than negative distress (Scheier, Weintraub, & Carver, 1986).

Optimists essentially expect to be successful at whatever they do, and they work harder than pessimists to reach their goals. As a result, optimists are beneficiaries of the **self-fulfilling prophecy** (Archibald, 1974). An optimistic mindset fuels a person's attempts to achieve more.

So, is optimism a *trait* determined by nature or a *state* determined by nurture or a particular environmental context or set of circumstances? The safest answer to this critical question is "Both." However, it is useful to realize that life circumstances can influence a transition from pessimism to optimism and vice versa.

While a personality trait predisposes individuals to be optimistic, pessimistic, or even cynical, the situation or context—especially behavioral consequences—certainly influences people to be more or less optimistic. Consider, for example, the overlap between feeling optimistic and feeling empowered and self-motivated (Safety Objective 3). Anything that fuels those person-states enhances an optimistic mindset.

Psychologists have distinguished between being optimistic—expecting the best—and being **hopeful**—working to achieve best expectations (Bryant & Cvengros, 2004). From this perspective, "I hope this happens," implies that "I can expect the best outcome if I do everything I know to do in order to achieve that outcome." Hope integrates optimism and personal control—invaluable

FIGURE 9.4 Type A is good or bad.

person-states for experiencing beneficial stress rather than debilitating distress.

TYPE A *BEHAVIOR* VS. TYPE A *EMOTION*

Figure 9.4 might seem familiar to many readers—perhaps too familiar. Many people have the mindset, "I have so much to do and not enough time to do it." In this state, personal goals are thwarted; personal control and optimism are diminished; and stress turns into distress. This mindset can elicit frustration, and frustration can lead to aggression and a demeanor that only increases distress. It can be a vicious cycle that certainly decreases the propensity for a healthy mindset, beneficial relationships, and/or injury prevention.

People with a **Type A disposition** are more likely to experience the time urgency and competitiveness depicted in Figure 9.4. The initial research of the Type A personality trait associated this disposition with a higher risk for coronary disease (Friedman & Ulmar, 1984), but subsequent research indicated that Type A plus hostility (as depicted in Figure 9.4) is the risk factor for a heart attack (Chida & Hamer, 2008; Chida & Steptoe, 2009). Thus, Type A safety professionals with an AC4P mindset are not more prone to coronary disease.

A particular personality scale can distinguish between Type A *behavior* and Type A *emotion* (Geller, 2008), with Type A *emotion* putting people at risk for heart disease. For example, while you are rushing to save time (behavior), do you feel anger toward those people who get in your way and slow you down (emotion)? Please consider our prior discussion of emotional intelligence or EQ (Safety Directive 11) and appreciate the need to replace negative emotions with a relevant positive mindset, like systems thinking in this case (Geller, 2023).

Type A *behavior* individuals are goal driven and self-motivated to get things done. They dislike wasting time and work hard to maintain personal control over life challenges they deem important. While these folks do experience more daily stressors than individuals with a **Type B disposition**, a success-seeking and hopeful perspective leads to the experience of stress rather than distress.

On the negative side, Type A individuals are more likely to experience frustration (Type A *emotion*), and this person-*state* can activate hostility and aggressive behavior. In addition, Type A folks often fail to appreciate the intrinsic value and flow of momentary pleasures, and they miss benefits from mindful meditation—relaxing inside and outside, and sensing the positive unique stimuli of the momentary present. Such EQ enables the suppression of debilitating or unhealthy Type A emotions.

MINDSET AND SELF-TALK

As discussed above and in prior safety objectives, a variety of environmental and interpersonal factors influence our mindset—from perceptions of personal control, optimism, and empowerment to feeling overwhelmed, frustrated, and distressed. Of course, cognitive processing or thinking precedes and follows a given mindset, both as a contributor and as an outcome of that mindset. Behavioral scientists define thinking as covert behavior or self-talk (Safety Objective 1).

Psychological science has demonstrated that intrapersonal conversations or self-talk about a task or challenge can facilitate or stifle success. This is commonsense to most readers because you have been there. You understand how self-talk can influence a success-seeking or a failure-avoiding mindset—a state of stress or distress.

The critical connection between self-talk and feeling empowered is obvious, right? We feel empowered when we believe (a) we can do the task; (b) the task is relevant to our mission or ultimate vision; and (c) completing the task is worth the effort. How do we acquire and sustain those three beliefs necessary to feel empowered?

Answer: Those three empowerment beliefs—self-efficacy, response efficacy, and outcome expectancy (Safety Objective 3)—are developed, nourished, and sustained through self-talk. Of course, our self-talk can also deny, disregard, or oppose those empowerment-determining beliefs. A failure-avoiding or a failure-accepting mindset can inhibit perceptions of personal control, competence, and a prophecy of success.

Consider also how our self-talk both influences and reflects our self-esteem. Actually, one could say that our mental script about ourselves is our self-esteem. We can focus our intrapersonal conversations on the good or bad things other people say about us. Our self-esteem goes up or down according to how we talk to ourselves about the way others talk about us (Safety Objective 1). Plus, the potential impact of the social media on our self-talk and our self-esteem is obvious to all readers.

THE LEGACY OF TEACHING AND LEARNING

Let's consider one final mindset we should all adopt and sustain—the remarkable legacy of teaching and learning. "We live, we love, we learn, and we leave a legacy." This profound quotation from Stephen R. Covey (1989) says it all. When people—our children, students, friends, and colleagues—learn from our teaching, they might not only apply the principles and procedures we teach, they might pass them on. In turn, they leave their own teaching/learning legacy.

Consider connections between Covey's Hierarchy of the four L-words and Abraham Maslow's Hierarchy of Needs (Maslow, 1971). At the bottom, we have the "to live" physiological needs. Next, we have the "to love" and "to learn" needs for social acceptance, self-esteem, and self-actualization. At the top of Maslow's Hierarchy is self-transcendence—the need to go beyond our own self-serving needs for someone else's wellbeing (Safety Objective 8). We call this "actively caring for people"—www.ac4p.org.

The connection between these need hierarchies proposed by two of the most provocative thought leaders of the 20th century is obvious, right? Maslow claimed the best we can be is to go beyond ourselves for others; Covey challenged us to leave a legacy of which we would be proud. Covey and Maslow have defined an enviable daily mission and long-term vision for us all. Practice the 60 safety directives recommended in this workbook to improve OHS and you exemplify self-transcendence. Teach others what you learn from applying these safety directives, and you will leave a positive teaching/learning legacy of which you can be extremely proud.

Conclusion

Safety Objective 9 targeted "mindset"—a human dynamic familiar to all readers and referenced frequently in various self-help books and by the public media as a determinant of human performance. We showed how a person's mindset—perception, belief, expectancy, or person-state—influences another popular term: "stress." Stress is not an undesirable person-state, as many seem to believe. Distress is the undesirable mindset, and the difference between positive stress and negative distress is the perception of personal control.

We introduced the person-state of optimism as a buffer for distress and explained how the psychological distinction between optimism and hope incorporates the mindset of personal control. Hoping for the best requires the application of relevant personal control in order to expect the best.

The Type A vs. Type B personality disposition connects directly to experiencing stress or distress. While Type A individuals experience more stressors than do Type B folks, a distinction between Type A behavior and Type A emotion is critical. The time urgency and competitiveness of Type A individuals result in stress and not distress if the negative Type A emotions of impatience, frustration, and aggression are minimized through intrapersonal conversation or self-talk. The propensity to display Type A behavior is essentially a personality *trait*, determined more by nature than by nurture, but Type A emotion is a person-state—a mindset—nurtured through personal and interpersonal intervention.

DISCUSSION QUESTIONS

1. Distinguish between "stress" and "distress" with a real-world example.

2. Explain the difference between "Domain of Influence" and "Domain of Concern" with regard to OHS.

3. Provide a distinction between "optimism" and "hope" with a real-world example.

4. Do you consider yourself to be Type A? Why or why not?

5. Please explain the distinction between Type A *behavior* and Type A *emotion* with reference to a safety-related situation. Why is this distinction important?

6. Do you consider Type A to be more of a personality *trait* or a personality *state*? Please explain.

7. Are you pleased that the top of Abraham Maslow's Hierarchy of Needs is "self-transcendence" and not "self-actualization"? Why or why not?

REFERENCES

American Heritage Dictionary. (1991). Second College Edition. New York: Houghton Mifflin Company.

Archibald, W. P. (1974). Alternative explanations for the self-fulfilling prophecy. *Psychological Bulletin, 81,* 74–84.

Bryant, F. B., & Cvengros, J. S. (2004). Distinguishing hope and optimism: Two sides of a coin, or two separate coins? *Journal of Social and Clinical Psychology, 23*(2), 273–302.

Carver, C. S., Scheier, M. F., & Weintraub, J. K. (1989). Assessing coping strategies: A theoretically-based approach. *Journal of Personality and Social Psychology, 56,* 267–283.

Chida, Y., & Hamer, M. (2008). Chronic psychosocial factors and acute physiological responses to laboratory-induced stress in healthy populations: A quantitative review of 30 years of investigation. *Psychological Bulletin, 134,* 829–885.

Chida, Y., & Steptoe, A. (2009). The association of anger and hostility with future coronary heart disease: A meta-analytic review of prospective evidence. *Journal of the American College of Cardiology, 17,* 936–946.

Covey, S. R. (1989). *The seven habits of highly effective people: Restoring the character ethic*. New York: Simon and Schuster.

Friedman, M., & Ulmar, D. (1984). *Treating Type A behavior and your heart.* New York: Knopf.

Geller, E. S. (2008). *Leading people-based safety: Enriching your culture* (Chapter 6). Virginia Beach, VA: Coastal Training Technologies Corporation.

Geller, E. S. (2023). Emotional intelligence: A crucial human dynamic for occupational safety and health. *Professional Safety, 68*(12), 24–29.

Maslow, A. H. (1971). *The farther reaches of human nature*. New York: Viking.

New Merriam-Webster Dictionary. (1989). Springfield, MA: Merriam-Webster, Inc., Publishers.

Scheier, M. F., Weintraub, J. K., & Carver, C. S. (1986). Coping with stress: Divergent strategies of optimists and pessimists. *Journal of Personality and Social Psychology, 51,* 1257–1264.

Selye, H. (1974). *Stress without distress.* Philadelphia, PA: Lippincott.

10 Safety Objective 10 *Cultivate and Sustain a Total Safety Culture*

Safety Directive 55: Treat safety as a value, not a priority.
Safety Directive 56: Promote safe behavior as a descriptive norm.
Safety Directive 57: Set safe examples all day and every day.
Safety Directive 58: Support safe behavior with sincere praise and gratitude.
Safety Directive 59: Teach the reciprocal benefits of expressing gratitude.
Safety Directive 60: Promote and support a safety suggestion system.

How can an organization achieve and sustain an injury-free workplace? Answer: "Cultivate and sustain a safety culture." That is the most popular quick-fix response we have heard for this critical question. Indeed, "safety culture" is the ultimate outcome or vision of any large-scale safety improvement process. Practically every presentation that addresses the prevention and/or reduction of workplace injuries on a large scale refers to "culture change," or the achievement of a "safety culture."

Scott introduced the vision of a Total Safety Culture (TSC) 30 years ago in his first book on the psychological science of safety (Geller, 1996), and we use the TSC label here. But what is a "TSC," and how can an organization achieve this vision? We address this challenge here for Safety Objective 10—our final safety objective of this occupational health and safety (OHS) teaching/learning workbook/manual.

WHAT IS A TSC?

The *American Heritage Dictionary* (1985) defines culture as "the attitudes, feelings, values, and behavior that categorize and inform society as a whole or any social group within it." (p. 348). Individuals in a TSC activate and support injury prevention by their routine behavior—from interpersonal conversations to active participation in safety-related interventions. With such safety-supportive behavior and related self-talk, they shift safety from a priority to a value—from a "safety-now" mindset influenced by current contextual demands to a "safety-always" mindset impervious to shifting priorities.

DOI: 10.1201/9781003758822-11

This paradigm shift to a TSC is easier said than done. In fact, every safety objective of this workbook explains evidence-based safety directives for safety leaders to implement on a regular basis in order to achieve and sustain a TSC—employees continually engaged in practicing the safety directives explicated in this workbook to achieve and maintain an injury-free workplace.

More specifically, safety leaders consider the complexity of human nature (Safety Objective 1) and then practice social influence principles (Safety Objective 5) and humanistic behaviorism (Safety Objective 8) when they (a) motivate safety-related behavior with extrinsic and intrinsic consequences (Safety Objective 2); (b) promote and support self-accountability for injury prevention and set empowering safety improvement goals (Safety Objective 3); (c) conduct an analysis of an injury or a close call (Safety Objective 4); (d) deliver behavior-based feedback to a coworker (Safety Objective 6); (e) contribute to attain interdependent trust and synergy (Safety Objective 7); and (f) develop a healthy mindset within oneself and among others (Safety Objective 9).

Yes, that lengthy sentence puts each safety objective of this workbook within the context of achieving and maintaining a TSC. With so many human dynamics referenced, the challenge of cultivating and sustaining a TSC can seem daunting. Here we attempt to simplify the TSC mission by recommending a few basic qualities of a TSC you can readily achieve with a few practical strategies. We begin by considering social norms. A primary characteristic of a TSC is that safe behavior is the norm—the behavior expected from everyone within an OHS-supportive context.

DESCRIPTIVE VS. INJUNCTIVE NORM

People gain information by observing the behavior of others. We are particularly observant of the behavior of others when we are in an unfamiliar setting. We watch what others are doing and saying in order to fit in. That is the power of observational learning or **social proof** as labeled by Robert Cialdini (2001). This is considered normative influence, and it defines a social norm. Social norms are injunctive or descriptive. An **injunctive norm** defines socially desirable behavior or what people "ought to do." Safe behavior is obviously injunctive, as is behavior that exemplifies actively caring for people (AC4P).

A **descriptive norm** is the commonly observed behavior of people in an organization or within a particular context. Injunctive norms are not necessarily descriptive, and as depicted in Figure 10.1, descriptive norms are not necessarily injunctive. In a TSC, safe behavior is both descriptive and injunctive.

FIGURE 10.1 Descriptive norms influence behavior.

It is noteworthy that behavior reflecting an injunctive norm can be activated by a descriptive norm. For example, guests at a hotel were most likely to reuse their towels to conserve energy when the message that requested them to hang up their used towels for reuse included the descriptive message that the prior guests who had stayed in their room reused their towels (Goldstein, Cialdini, & Griskevicius, 2008).

Thus, safety leaders can influence more occurrences of a particular safe behavior by informing employees that their coworkers consistently perform that behavior (e.g., "Your work team always locks out the power before adjusting that energized equipment."). Of course, such social influence statements are true in a TSC where descriptive norms are injunctive and vice versa.

SET THE SAFE EXAMPLE

Let's consider the most basic norm-influencing process of psychological science—observational learning. If you want to be better at what you do, watch someone who performs that task better than you do. The power of observational learning is obvious; a large body of psychological research indicates that this type of learning is part of almost everything we do (Bandura, 1969).

People's actions are the result of observational learning whenever they do something in a particular way. They have seen somebody else do it that way,

FIGURE 10.2 We learn much from observation.

or someone shows them how to do it that way, or characters on television or in a video game do it that way. However, as illustrated in Figure 10.2, we often do not realize the observational learning influence we have on others.

Employees learn numerous patterns of job behaviors by watching their coworkers. When they see a coworker receive positive recognition for a certain behavior, they are more likely to perform that behavior. This behavioral influence is termed **vicarious** *reinforcement* (Bandura, 1969). At the same time, when employees observe others receive a negative consequence for performing certain work behavior, they learn to avoid or stop that behavior. Bandura (1969) referred to this behavioral influence as **vicarious** *punishment*.

The occurrence of safe behavior (e.g., using PPE and fall protection) encourages similar behavior by observers, and the concomitant verbal behavior can be influential. If a worker observes a supervisor commending or reprimanding another worker for his/her safe or at-risk behavior respectively, the observer might be influenced to increase his/her performance of safe behavior through vicarious reinforcement or to decrease the frequency of an at-risk behavior through vicarious punishment.

Bottom line: To make safe behavior the norm–rather than the exception–employees in a TSC set the safe example both in their own practices and in their communication with others. You never know when someone is observing and learning from your behavior. If you truly believe in a TSC, you will attempt to consistently walk the safety talk and vice versa. The power of observational learning is exemplified by the following poem by Forrest H. Kirkpatrick.

> The eye's a better teacher and more willing than the ear;
>
> Fine counsel is confusing, but example's always clear;
>
> And the best of all the preachers are the ones who live their creeds;
>
> For to see the good in action is what everybody needs.
>
> I can soon learn how to do it if you let me see it done;
>
> I can watch your hands in action, but your tongue too fast may run;
>
> And the lectures you deliver may be very wise and true;
>
> But I'd rather get my lesson by watching what you do.
>
> For I may not understand you and the high advice you give;
>
> There's no misunderstanding how you act and how you live.

PROVIDE BEHAVIORAL SUPPORT

The most efficient and effective behavior employees could perform on a regular basis to activate and support a TSC—offer sincere one-to-one praise or gratitude for observed behavior that reflects a TSC. Yet, as illustrated in Figure 10.3, words of approval, appreciation, and praise for commendable behavior are relatively rare, especially when compared to the use of verbal reprimands or correction for undesirable behavior.

As reflected in Safety Directives 33 and 34, learning is more effective and enjoyable when success is rewarded more often than errors are penalized. Errors are disruptive behaviors that stick out and invite attention and corrective feedback, but desirable behavior does not naturally attract attention and seemingly does not require intervention such as behavioral recognition or an expression of gratitude.

FIGURE 10.3 Positive recognition is too rare.

Both research and commonsense tell us it is extremely beneficial to call attention to behaviors we want to occur more often and to recognize those desirable behaviors appropriately. Behavioral scientists refer to these positive consequences as *now-that* rewards (Safety Directive 8), in contrast to *if-then* rewards or incentives. As explained for Safety Objective 2, the delivery of now-that recognition is simple and straightforward. You simply specify the desirable safe behavior observed and then offer sincere praise, recognition, and/or gratitude.

The result: You might increase occurrences of the recognized safe behavior, you will surely enhance the person's perception of personal competence and self-motivation. Behavioral praise not only enhances self-esteem, ~~but~~ it fuels a perception of competence at performing certain desirable behaviors. Psychologists call this self-efficacy (Bandura, 1997), and that person-state is key to being self-motivated and feeling empowered (Safety Objective 3).

THE POWER OF GRATITUDE

Substantial research indicates that gratitude—the person-state of feeling grateful—significantly increases **subjective wellbeing (SWB)** or life satisfaction (e.g., Emmons & Crumpler, 2000; Wood, Froh, & Geraghty, 2010). More specifically, gratitude has been shown to enhance positive emotions and to activate a sense of interpersonal belonging, while decreasing distress (Safety Objective 9) and depression (Emmons, 2007; Emmons &

McCullough, 2003). In fact, people are more likely to help others—perform AC4P behavior—when they feel grateful (Emmons & Mishra, 2011).

So, how can we increase perceptions of gratitude and experience the beneficial side effects of this person-state? You know the answer—offer a sincere statement of personal recognition for another individual's AC4P behavior. Indeed, when you thank someone for the performance of safety-related or health-related behavior you are expressing gratitude for the support of a TSC. Plus, expressing gratitude for another person's effort to keep others safe reflects the achievement of self-transcendence—the top of Maslow's revised Hierarchy of Needs (Safety Directive 48).

Surveys have found that people are less likely to express gratitude at work than at any other place. In fact, 60% of more than 2,000 respondents in online interviews reported that they never or very rarely thank anyone at work; only 10% said they express interpersonal gratitude at work on a given day, and 74% said they never or rarely express gratitude to their boss or supervisor (Kaplan, 2012).

Relatedly, Krista administered a safety culture survey and several focus group discussions at four facilities of a major manufacturing company. A key conclusion from 615 survey respondents and 198 focus group participants: Positive recognition for safety-related behavior was rare at each of the four plants, and the employees stated a sincere desire to receive positive feedback for their safety-relevant efforts (Geller, 2020).

A Reciprocal Benefit

Who experiences a boost in SWB when one person thanks another for desirable behavior s/he observes? Obviously, the individual receiving the recognition appreciates the positive interpersonal exchange and likely experiences an increase in SWB, competence, and self-motivation (Safety Objective 3), and feels a positive connection with the benefactor—the person who expressed safety-related gratitude.

How does the expression of gratitude affect the benefactor? You know the answer because you have been there and have experienced the reality of the expression, "It's better to give than to receive." Giving recognition or showing appreciation enhances one's person-state of gratitude and therefore his/her SWB. For example, seminal research by Martin Seligman—the initiator of the positive psychology movement in 1998—and his colleagues demonstrated a most powerful way to increase personal gratitude and

SWB: Write a thank-you letter to a benefactor and read it to that individual (Seligman et al., 2005).

THE VALUE OF SAFETY SUGGESTIONS

Given the numerous factors that influence the human dynamics of an organizational culture, a mission to develop and nurture a TSC can feel overwhelming—eliciting detrimental distress rather than beneficial stress (Safety Directive 49). Let's break down this challenge to the promotion and support of just one critical behavior required for a TSC.

What behavior would you choose to promote and support at your workplace to enable the achievement of a TSC and contribute to making safety-related behavior a descriptive norm? A group discussion of various answers to this crucial question would likely be interesting, informative, and inspirational. Our answer to this question—the daily submission of safety improvement suggestions from line workers.

Who knows better how to make a workplace safer than those employees on the front line? Almost every day these workers perceive safety hazards, experience close calls directly or indirectly, observe at-risk behavior, and periodically envision ways to make a job safer. How often do these line workers translate their safety-related experiences into a suggestion for improving safety management and cultivating a TSC? Does Figure 10.4 ring true for any workplace you have experienced?

FIGURE 10.4 Critical behaviors need support.

Do you have a "safety suggestion box" at your workplace in which employees can place safety suggestions anonymously? If so, how many safety suggestions do you receive daily, and what percentage of those suggestions result in a beneficial change in the environment, safety-related policy, or the safety management system? To what extent would such descriptive statistics indicate whether the organization has achieved a TSC?

Given email and text messaging, a safety suggestion box might be considered unnecessary today. However, anonymity can be difficult with digital communication, and there could be an advantage to having visible safety suggestion boxes on the "shop floor," with an opportunity to offer immediate on-the-job comments. For a safety suggestion system to work effectively, someone needs to provide soon and certain feedback for every suggestion—if only to express gratitude for a suggestion that cannot be implemented for a specified rationale.

Many years ago, Scott worked with safety leaders at a Toyota motor manufacturing plant in Georgetown, Kentucky, whose 6,000 employees submitted more than 35,000 quality, production, or safety-related suggestions in 1994. Many employees in this Kaizen (or continuous improvement) culture were self-motivated to submit suggestions, but extrinsic consequences were also in place to motivate participation.

The employees received timely feedback regarding the utility and feasibility of every suggestion; and if a suggestion was approved, the relevant employees were empowered to implement it themselves whenever possible. Moreover, the individual or team responsible for the suggestion received 10% of any financial savings or profits that resulted from the implemented suggestion for the first year. At the time, this suggestion system involved three full-time employees evaluating and providing feedback for each suggestion and cost $1 million per year. However, each year this continuous improvement process reaped $5 million in estimated savings and/or profits (Geller, 1996).

Conclusion

This final safety objective of our workbook on applications of psychological science to achieve and sustain an injury-free workplace considered the most popular "buzz word" among both safety professionals and consultants: "culture." We addressed the cultivation of a TSC by referring to each of our prior contributions to this teaching/learning safety handbook/manual. Then we selected a few primary and practical techniques companies need to adopt and adapt in order to promote and support a TSC.

Specifically, to make safety-related behavior a descriptive norm, it is necessary to (a) recognize the powerful influence of observational learning and consistently set safe examples for others; (b) provide routine support for coworkers' safe behavior by delivering sincere appreciation and gratitude for their AC4P behavior; (c) promote and support a safety suggestion system that encourages employees to submit suggestions that could decrease the frequency of injuries by changing environmental, management system, and/or behavioral factors; and (d) implement a safety suggestion system that provides soon and certain feedback per each suggestion received and applies those suggestions that are feasible and injury-preventive.

We hope it is obvious that the key to cultivating and sustaining a TSC is comprehensive employee engagement—all employees applying their on-the-job experience to identify ways to make behavior and its context safer. This includes setting safe examples daily through their own work behavior and by supporting the safe behavior of their coworkers with interpersonal expressions of gratitude for their AC4P efforts. In a TSC, behavior that promotes and supports safety and AC4P is a descriptive norm.

DISCUSSION QUESTIONS

1. Should safety be a "priority" or a "value"? Please explain your answer with a workplace example.

2. Explain the distinction between a descriptive norm and an injunctive norm with real-world examples.

3. What safe and at-risk behaviors have you learned by observing the behavior of others?

4. Provide a practical example of using vicarious reinforcement to increase the occurrence of a particular safe behavior at your work site.

5. Describe the last time you received gratitude–a sincere thank you from another person–for something you did. How did that interaction make you feel?

6. Describe the last time you delivered gratitude for another individual's AC4P behavior.

7. Why are interpersonal expressions of gratitude relatively rare in the workplace?

8. How does your workplace solicit safety suggestions from employees?

9. How could the safety suggestion system at your workplace be improved for greater employee engagement and beneficial applications of the recommendations?

10. From your experience, what are the most important qualities of a TSC?

REFERENCES

American Heritage Dictionary. (1985). Second College Edition. Boston, MA: Houghton Mifflin Company.

Bandura, A. (1969). *Principles of behavior modification*. New York: Holt, Reinhold, & Winston.

Bandura, A. (1997). *Self-efficacy: The exercise of control*. New York: Freeman.

Cialdini, R. B. (2001). *Influence: Science and practice* (6th ed.). Boston, MA: Pearson.

Emmons, R. A. (2007). Gratitude, subjective well-being, and the brain. In R. J. Larsen & M. E. McCullough (Eds.), *The science of subjective well-being*. New York: The Guilford Press.

Emmons, R. A., & Crumpler, C. A. (2000). Gratitude as a human strength: Appraising the evidence. *Journal of Social and Clinical Psychology*, *19*(1), 56–69.

Emmons, R. A., & McCullough, M. E. (2003). Counting blessings versus burdens: Experimental studies of gratitude and subjective well-being in daily life. *Journal of Personality and Social Psychology, 84*, 377–389.

Emmons, R. A., & Mishra, A. (2011). Why gratitude enhances well-being: What we know, what we need to know. In K. M. Sheldon, T. B. Kashdan, & M. F. Steger (Eds.), *Series in positive psychology. Designing positive psychology: Taking stock and moving forward* (pp. 248–262). New York: Oxford University Press.

Geller, E. S. (1996). *The psychology of safety: How to improve behaviors and attitudes on the job*. Boca Raton, FL: CRC Press.

Geller, K. S. (2020, January). *Final qualitative and quantitative data report*. Newport, VA: GellerAC4P, Inc.

Goldstein, N. J., Cialdini, R. B., & Griskevicius, V. (2008). A room with a viewpoint: Using social norms to motivate environmental conservation in hotels. *Journal of Consumer Research*, *35*(3), 472–482.

Kaplan, J. (2012). *Gratitude Survey by the John Templeton Foundation*. West Conshohocken, PA.

Seligman, M. E. P., Steen, T. A., Park, N., & Peterson, C. (2005). Positive psychology progress: Empirical validation of interventions. *American Psychologist*, *60*(5), 410–421.

Wood, A. M., Froh, J. J, & Geraghty, A. W. A. (2010). Gratitude and well-being: A review and theoretical integration. *Clinical Psychology Review, 30*(7), 890–905.

Glossary of Key Terms

ABC (activator-behavior-consequence): the three-term contingency of applied behavioral science (ABS), specifying that a stimulus event (an activator) occurring before a behavior provides direction (e.g., an incentive or a disincentive), whereas the consequence following a behavior provides motivation to repeat or stop the preceding behavior.

Activator: an environmental event (e.g., directions from a sign or a person, an incentive or disincentive) implemented to influence the occurrence of a particular target behavior.

Actively caring for people (AC4P): the application of behavioral science and select principles from humanism to increase the frequency and/or improve the quality of behavior that benefits human health, safety, wellbeing, and/or welfare—referred to as humanistic behaviorism.

AC4P coaching: interpersonal communication whereby one individual (the teacher, coach, or parent) employs principles of humanistic behaviorism to benefit the observed behavior of another individual (the learner).

AC4P culture: an environmental context or setting (e.g., a workplace, school, or home) where people interact daily on behalf of the health, safety, security, and wellbeing of everyone else in the surroundings with a spirit of win/win interdependence and self-transcendence.

Applied behavioral science (ABS): the application of research-based principles derived from experimental behavior analysis to increase the occurrence of desirable behaviors and decrease the frequency of undesirable behaviors.

Attributional bias: a prejudiced assumption of why a person is acting in a certain way—reactive thinking as opposed to reflective thinking.

Authority principle: the tendency to comply with a request from a person with presumed credibility and perhaps influence over relevant behavioral consequences.

Behavior-based feedback: interpersonal communication following the occurrence of an observed behavior that shows the performer what s/he did correctly (supportive feedback) and/or incorrectly (corrective feedback).

Behavior-based safety (BBS): the systematic application of ABS to prevent personal injury by increasing the occurrence of safe behavior and decreasing the frequency of at-risk (or unsafe) behavior, typically manifested by a peer-to-peer behavior observation-and-feedback process.

Bystander apathy effect: the tendency for any given observer of a person needing help to be less likely to give aid if other bystanders are present, presumably due to diffusion of personal responsibility.

Choice: the perception of having more than one option with regard to accomplishing a particular task or action plan.

Classical conditioning: a learning process whereby a previously neutral stimulus comes to elicit (or cause) an automatic or reactive response (or behavior)—an involuntary conditioned response.

Cognitive dissonance: tension or discomfort experienced when an individual's behavior is inconsistent with his/her attitude, belief, and/or values, or vice versa.

Community: the perception of relatedness or belongingness with others, and often an experience of interdependent social or group support regarding the accomplishment of a certain assignment or challenge.

Compassion: AC4P action or behavior following empathy—a sincere concern for another person's welfare or wellbeing.

Competence: the perception or belief that an individual has the knowledge, skill, and ability to accomplish a certain task effectively.

Conditioned Stimulus: a stimulus previously neutral with regard to triggering a particular response acquires the capacity to elicit that same response (a conditioned response) through the learning process of classical conditioning.

Consequence: an environmental event (positive or negative) that naturally occurs (i.e., intrinsic) after a behavior or is added to the situation (i.e., extrinsic) in order to influence a certain behavior, which may or may not influence the frequency or improve the quality of the preceding behavior.

Consistency: the pressure to behave in ways that are in concert with our attitudes, beliefs, or values, or to behave in ways others expect us to behave.

Contingency management: manipulating an activator and the consequence(s) of a target behavior in order to increase (with an incentive/reward) or decrease (with a disincentive/penalty) its quality or its frequency of occurrence.

Corrective feedback: behavior-focused information given to an individual after s/he performs an undesirable or less-than-optimal behavior, often manifested in an one-to-one coaching interaction.

Countercontrol: the occurrence of a behavior that is contrary to a desired target behavior because the person perceives a loss of personal choice or individual freedom as the result of a top-down rule, a disincentive, or an if-then contingency.

Defensive pessimism: a strategy used by failure avoiders to help them manage their anxiety so they can work productively. They lower their expectations in order to help prepare themselves for the worst.

Descriptive norm: the commonly observed or reported behavior of other people.

Discretionary behavior: desirable behavior that was not required by an extrinsic accountability system—presumed to be self-directed and self-motivated.

Disincentive: an activator (or extrinsic stimulus) that announces the delivery of a certain negative consequence (a penalty) following the occurrence of a designated unwanted behavior.

Disposition: an internal trait or person-state of an individual that predisposes or influences his/her behavior at a certain time and place.

Distress: the physiological and psychological reaction to an event that is perceived as important—a stressor—but appraised as threatening or overwhelming and beyond the person's current domain of personal control.

Door-in-the-face technique: making a large, irrational, behavioral request in order to make a subsequent smaller request seems much more reasonable and doable.

Education: teaching one or more individuals awareness, theory, or procedures related to a certain task or the circumstances related to that task.

Emotion: a potent and potentially long-term reaction to a person or a situation that can motivate relevant self-directed behavior.

Emotional intelligence (EQ): one's ability to (a) remain in control and optimistic following personal failure and frustration; (b) understand and empathize with other people and work with them cooperatively; and (c) give up a soon, certain, and positive consequence for a delayed and uncertain but larger positive consequence.

Empathic listening: the highest level of interpersonal listening whereby the listener attempts to identify with and understand the speaker's feelings, attitude, motives, and behavior, as well as the circumstances influencing relevant dispositions and behavior.

Empathy: attempting to identify with and understand another individual's circumstances that influence his/her feelings, attitude, motives, and behavior.

Empowerment: an individual's belief or confidence that s/he can master a particular assignment or accomplish a certain task, determined by answering "Yes" to three questions—Can I do it? Will it work? and Is it worth it?

Extrinsic: a stimulus event external to the individual, considered to be an activator or a consequence when applied to influence (or motivate) the occurrence of a target behavior.

Extrinsic motivator: an activator (e.g., an incentive or disincentive) that influences the occurrence of a behavior in order to gain a positive consequence, avoid a negative consequence, or escape an aversive environmental event.

Extrinsic reward: a positive behavioral consequence delivered after the occurrence or outcome of behavior(s) in order to support the behavior(s) and/or to enhance an individual's sense of competence or self-efficacy.

Extrinsic stimulus: an observable and external environmental event that might direct behavior (as an activator) or motivate behavior (as a positive or negative consequence).

Facial feedback technique: manipulating one's facial muscles (e.g., force a smile) to trigger a corresponding emotion (e.g., happiness or positive affect).

Failure avoider: an individual performing a certain behavior in order to avoid a negative consequence.

Feedback: behavior-focused information (supportive or corrective) following the performance of a behavior, delivered to influence the frequency and/or the quality of the preceding behavior.

Foot-in-the-door technique: soliciting a small behavioral commitment for a person to perform a relatively convenient behavior in order to obtain subsequent compliance with a more inconvenient behavior with greater response cost.

Fundamental Attribution Error: the tendency for an observer of another person's behavior to underestimate the impact of the context (situational factors) and overestimate the impact of dispositional factors (person-states and/or traits).

Gratitude: an expression or a reception of sincere appreciation for an individual's behavior that temporarily boosts the positive person-state of subjective wellbeing (SWB) for both the benefactor and the beneficiary.

Groupthink: the tendency for group members to agree with a person of authority, resulting in conformity from individuals within that group who may hold a different view but do not express dissention.

Hope: the combination of optimism and personal control, contributing to the belief that positive expectations will eventuate if relevant behaviors are performed.

Humanism: an appreciation of empathy, diversity, and human dignity, and a sincere concern for human wellbeing.

Humanistic behaviorism: making applications of ABS implemented to improve human performance more effective by practicing empathy and considering such person-states as self-esteem, self-efficacy, belongingness, personal control, and optimism.

Hypocrisy effect: the occurrence of a desired target behavior is motivated by activating a "guilt trip"—participants perceive an inconsistency between their current personal commitment to perform a certain desirable behavior and their lack of previously performing that behavior (i.e., they realize an inconsistency between their prior behavior and a current behavioral commitment).

If-then reward: a positive behavioral consequence whereby an activator had announced the delivery of a positive consequence following the occurrence of a certain desired behavior—an if-then reward contingency.

Incentive: an announcement (an activator) that specifies the availability of a positive consequence (a reward) following the performance of a designated behavior or a desirable outcome of more than one behavior.

Independence: an individualistic viewpoint or mindset that sets the occasion for self-focused behavior ("I can do this myself") and inhibits other-focused behavior ("Let's work together to make this happen").

Injunctive norm: the desirable behavior in a particular situation or what one "ought to do."

Interdependence: a collectivistic, we-need-each-other viewpoint or mindset that facilitates and supports AC4P behavior—acts performed on behalf of the health, safety, welfare, and/or the wellbeing of others.

Interpersonal trust: the AC4P connection between people that enables the cultivation of an AC4P culture with the seven C's—communication, caring, candor, consistency, commitment, consensus, and character—along with the interpersonal behavior-based feedback tools provided in this OHS handbook.

Intervention: an external or extrinsic program or process implemented to influence the quantity or quality of one or more target behaviors.

Intrinsic consequence: a positive or negative stimulus event that flows naturally after a behavior and keeps it going (a positive reinforcer) or causes it to decrease in frequency or stop completely (a punisher).

Leadership: inspiring people to be self-motivated and feel empowered to perform one or more goal-directed behaviors (i.e., discretionary behavior).

Management: motivating one or more behaviors to occur through an external or extrinsic accountability system (e.g., activator, behavioral feedback, incentive, disincentive, rule, mandate, interpersonal recognition, or reprimand).

Nondirective: practicing empathic listening to learn another person's perceptions and relevant circumstances before providing advice or direction.

Now-that reward: a positive behavioral consequence (e.g., behavior-based recognition or supportive feedback) delivered after observing the occurrence of a desirable behavior—given to express sincere appreciation for the behavior and to potentially influence the recipient's perceived competence and self-motivation, as well as to increase the probability that the desired behavior will reoccur.

Observational learning: initiating or improving the performance of one or more behaviors by watching one or more individuals perform the behavior(s).

Optimism: the expectation that the consequences of one or more behaviors will be positive—a challenge will be handled successfully.

Outcome expectancy: anticipating the positive consequence(s) to follow the occurrence of one or more goal-directed behaviors and believing that achieving the consequence is worth the effort and time involved—the response cost.

Participative management: empowering people to help design and/or implement the contingencies or the accountability system that controls one or more of their own behaviors.

Penalty: a negative consequence following a designated behavior or an outcome of one or more behaviors that was intended to decrease recurrence of the behavior(s).

Perceived choice: the perception of having more than one option with regard to accomplishing a particular task or action plan.

Perceived competence: the perception or belief that an individual has the knowledge, skill, and ability to accomplish a certain task effectively.

Performance: the output or outcome of a process or system which includes input from behavior, dispositions (person-states and traits), and situational factors.

Personal control: one's perception of having the ability or competence to accomplish a particular task successfully—feeling empowered to meet a challenge.

Person-state: a personal disposition (e.g., expectancy, attitude, or personality characteristic) that influences an individual's behavior, but varies as a function of the situation or environmental context.

Positive reinforcement: a procedure or process that increases the frequency or improves the form of a behavior by following that behavior with a desirable consequence–termed a "positive reinforcer."

Premature cognitive commitment: the notion that people often hold on stubbornly to a preconceived perception about someone or something (i.e., inflexible prejudice) as the result of prior personal experience, attitudes, cognitions, and/or behaviors.

Principle of Authority: the tendency to comply with a request from a person with presumed credibility and perhaps has influence over relevant behavioral consequences.

Principle of Consistency: the research-based dynamic that people alter their attitudes, beliefs, and/or perceptions to be consistent with their behavior and vice versa.

Principle of Reciprocity: the social norm that obligates people to repay others with similar behavior (positive or negative) they have received from them—"pay-it-forward" or "pay back."

Principle of Scarcity: the research-based dynamic that the value of something—from a commodity to a perception of individual freedom, personal control, or quality of life—increases when its availability is challenged or appears rare or limited.

Psychological reactance: the perception of a loss of personal choice or freedom due to a top-down rule, disincentive, or an if-then contingency that activates the performance of behavior contrary to the desired behavior targeted by the rule, disincentive, or an if-then contingency.

Psychological safety: an important human dynamic whereby an individual feels accepted and included in a particular organizational culture, and believes s/he will be respected when speaking up with questions, concerns, mistakes, or new ideas.

Reinforcer: an extrinsic (extra) or intrinsic (natural) stimulus or environmental event that increases the frequency and/or the form of the behavior it follows.

Response efficacy: an individual's belief or confidence that performing one or more goal-directed behaviors will contribute to achieving a desired outcome, mission, or long-term vision.

Reward: a positive consequence following a designated behavior or an outcome of more than one behavior that may or may not influence recurrence of the behavior(s).

Selection by consequences: one of B.F. Skinner's legacies that reflects the motivational principle that people do what they do to gain a positive consequence or to avoid or escape a negative consequence.

Selective perception: the process of interpreting and organizing sensory information that is biased by prior experience and/or expectations and may influence attitude, thinking, and behavior.

Self-accountability: internal motivation (from within an individual) to perform one or more self-directed behaviors.

Self-actualization: the top of Maslow's *initial* Hierarchy of Needs at which a person feels a sense of ultimate achievement—having fulfilled one's potential.

Self-directed behavior: self-motivated behavior that is not solely motivated by an external or extrinsic motivator (e.g., an incentive or a disincentive).

Self-efficacy: an individual's belief or confidence that s/he has the knowledge, skill, and ability to perform a goal-directed behavior needed to accomplish a certain task.

Self-esteem: a general or overall feeling of self-worth that influences one's propensity to perform AC4P behavior.

Self-fulfilling prophecy: a prediction or expectation that directly or indirectly becomes true because relevant behavior is influenced to confirm the prophecy.

Self-motivation: a person-state that reflects internal drive to perform a certain behavior or achieve a particular outcome of one or more behaviors from a self-directed and self-accountability perspective.

Self-Serving Bias: taking personal credit for our successes, but blaming our failures on situational factors in order to maintain a positive view of ourselves and protect our self-esteem.

Self-talk: intrapersonal conversation (or covert behavior) that can influence overt behavior, and often reflects and/or affects one's attitude, perception, self-esteem, and/or self-motivation.

Self-transcendence: the top of Maslow's *revised* Hierarchy of Needs whereby an AC4P mindset is realized, and the individual experiences personal fulfillment and positive reinforcement when performing AC4P behavior—behavior contributing to the health, safety, welfare, or wellbeing of one or more persons.

SMARTS goal: a behavior-focused achievement objective defined by making it **S**pecific, **M**otivational, **A**ttainable, **R**elevant, **T**rackable, and **S**hared.

Social label: a positive or negative attribute given overtly to an individual which influences that individual to perform behavior consistent with that label.

Social proof: the tendency to look to others for guidance on what behaviors are appropriate or inappropriate in a particular situation, especially in an unfamiliar setting.

Stereotyping: assuming (or prejudging) that members of a particular group share certain characteristics or behaviors.

Stress: the physiological and psychological reaction to an event perceived as important and challenging—a stressor—and appraised as within one's current domain of personal control and empowerment.

Subjective wellbeing (SWB): a positive person-state of perceived happiness or life satisfaction influenced by interpersonal and intrapersonal conversation, among other contributing factors.

Success seeker: an individual performing one or more behaviors in order to earn a positive consequence, which can be intrinsic or extrinsic.

Supportive feedback: a now-that reward (or positive recognition) delivered after observing an individual's desirable behavior, which is most rewarding and influential when it can SOAR—it is *Specific* (designates the behavior); ***On** time* (occurs soon after the behavior is observed); *Appropriate* for the knowledge, ability, and experience of the performer; and shows ***Real*** appreciation for the individual's competence and effort.

Synergy: a process of interdependent participation in a goal-directed activity that results in greater achievement than possible from all participants performing alone and independently.

Systems thinking: a perspective of win/win, interdependency, teamwork, and synergy that optimizes an organizational, educational, or family system; realizing that the consequence for one behavior can serve as the activator for the next behavior, leading to a spiral causality of AC4P-focused synergism initiated by one AC4P behavior and the Principles of Reciprocity and Consistency.

Training: providing behavior-focused direction and feedback regarding the quality of one or more specified behaviors performed to achieve a designated outcome.

Trait: a disposition to feel and act in a particular way, presumed to be a stable lifelong personality characteristic resulting from nature rather than nurture.

Type A personality: a disposition (or person-trait) reflecting people who are competitive, ambitious, hard-driving, impatient, aggressive, and anger-prone.

Type B personality: a disposition (or person-trait) reflecting people who are easygoing, relaxed, and laid back—capable of living in the present and enjoying the moment.

Vicarious punishment: observational learning whereby the negative consequence given to one or more persons following undesirable behavior influences an observer of this interaction to decrease the occurrence of that behavior.

Vicarious reinforcement: observational learning whereby the positive consequence given to one or more persons following a desirable behavior influences an observer of this interaction to increase the frequency or improve the quality of his/her performance of that behavior.

Vision: a distant or ultimate objective or aspiration, presumed to be reached or achieved through ongoing SMARTS goal setting and the successive achievement of the relevant SMARTS goals.

Index

F

G

H

I

K

L

M

N

O

P

For Product Safety Concerns and Information please contact our EU representative GPSR@taylorandfrancis.com Taylor & Francis Verlag GmbH, Kaufingerstraße 24, 80331 München, Germany